The Best American Science and Nature Writing 2009

GUEST EDITORS OF
THE BEST AMERICAN SCIENCE
AND NATURE WRITING

2000 DAVID QUAMMEN
2001 EDWARD O. WILSON
2002 NATALIE ANGIER
2003 RICHARD DAWKINS
2004 STEVEN PINKER
2005 JONATHAN WEINER
2006 BRIAN GREENE
2007 RICHARD PRESTON
2008 JEROME GROOPMAN
2009 ELIZABETH KOLBERT

The Best American Science and Nature Writing™ 2009

Edited and with an Introduction
by Elizabeth Kolbert

Tim Folger, Series Editor

A Mariner Original

HOUGHTON MIFFLIN HARCOURT

BOSTON · NEW YORK   2009

The Best American Series® is a registered trademark of Houghton Mifflin Harcourt Publishing Company. *The Best American Science and Nature Writing*™ is a trademark of Houghton Mifflin Harcourt Publishing Company.

www.hmhbooks.com

ISSN 1530-1508

ISBN 978-0-547-00259-0

Printed in the United States of America

DOC 10 9 8 7 6 5 4 3 2 1

Contents

Contents

Foreword

WHAT SECRET INGREDIENTS make for a compelling story about science? One editor I worked with some years ago claimed to know what at least two of them were and never hesitated to mention their absence in one of my drafts. "What this story needs," he would say, "is some awe and wonder." I always cringed when I heard the advice, not because it was wrong but because it was so ineffable. Most journalists can handle basic clarity and structure — the who, what, when, where, and why. But awe and wonder? How does one get them into a story?

The challenge is especially daunting for writers who tackle scientific subjects, because jargon can obscure astonishing achievements. While writing an article about the nature of time a few years ago, I interviewed a physicist who told me that he could measure the most fleeting subatomic events, phenomena that lasted for just 100 attoseconds. Now an attosecond is 1 quintillionth of a second — that's a billionth of a billionth of a second. Impressive. But the true significance of the physicist's accomplishment didn't really sink in for me until I worked out that 1 second is to 100 attoseconds as 300 million years is to 1 second. In the attosecond world, a blink would last an eon.

Sometimes the bare facts of a story speak for themselves, with the awe and wonder immediately apparent. Last fall NASA scientists reported that they had detected snow falling from clouds on Mars, the first extraterrestrial snowfall ever observed, measured by an instrument aboard a NASA probe that landed on a red point of light more than 30 million miles from Earth.

But I think some of the best stories emerge when events so seize a writer that he or she can't help communicating a sense of amazement, passion, or tragedy. There's an urge to share; cords of concern, empathy, and interest bind reader, writer, and story. Chris Carroll's "High-Tech Trash" and David Quammen's "Contagious Cancer," both included in this collection, seem to cry out, "You need to look at this!"

Perhaps there's another secret ingredient in such exceptional stories. The gravestone of Arthur C. Clarke, the visionary science fiction writer who died in 2008 at the age of ninety, bears this epitaph, at his request: "Here lies Arthur C. Clarke. He never grew up and did not stop growing." Maybe it was a lifetime of contact with science — a discipline that requires the questioning of received wisdom — that infected Clarke's spirit. That same spirit is amply displayed within these pages.

For a nice dose of awe, and an unsurpassed example of scientific curiosity pushed to the limits, read Joshua Roebke's "Reality Tests." Roebke writes about a team of physicists who have set up an experiment to question one of our most fundamental shared assumptions about the world: namely, that it exists independently of our observations. Roebke no doubt worked very hard to extract the wonder of his story from the raw materials of interviews laden with the arcane language of theoretical physics.

Or turn to Virginia Morell's "Minds of Their Own," which doesn't skimp on wonder either. In the course of her story, Morell meets with a number of scientists who study animal intelligence. But the most memorable characters in the article are the animals being studied, like Rico, a border collie who recognizes the names of two hundred different toys, or Alex, an African gray parrot who didn't just parrot phrases but learned to use the English language creatively. Morell's article offers convincing evidence that intelligence, self-awareness, and the ability to understand grammar are not traits unique to humans. She writes that Charles Darwin believed that even earthworms displayed signs of intelligence.

This year marks the two hundredth anniversary of Darwin's birth, and the one hundred and fiftieth anniversary of the publication of *On the Origin of Species*. So it seems especially fitting that this collection includes "Darwin and the Meaning of Flowers," a beautiful essay by Oliver Sacks on the inspiration and delight Darwin found in his lifelong study of botany. Sacks himself shares Darwin's fascination with botany, and he concludes his essay with a remarkable paean to the interrelatedness of all living things.

The other twenty-one stories in the book are equally absorbing, from Atul Gawande's fascinating — and often disturbing — account of the daunting complexity behind the seemingly simple phenomenon of an itch, to Stephen S. Hall's article about the disappearance of the Neanderthals. In a way this book, like many good books, is like the mysterious black monolith in Arthur C. Clarke's novel *2001: A Space Odyssey*. The monolith transmitted knowledge to anyone who touched it. So will this book. Open it, read it, and be changed. You might not be inspired to wield a new tool, like the primates in Clarke's novel, but perhaps the world will look a bit different, a bit richer, a bit less familiar. This collection will certainly transform ordinary hours into ones that will be by turns (of the page) thoughtful, mysterious, sad, and, of course, awesome and wonderful. I think my editor friend would be pleased.

I hope that readers, writers, and editors will nominate their favorite articles for next year's anthology at http://timfolger.net /forums. The criteria for submissions and deadlines, and the address to which entries should be sent, can be found in the "news and announcements" forum on my Web site. I also encourage readers to use the forums to leave feedback about the collection and to discuss all things scientific. The best way for publications to guarantee that their articles are considered for inclusion in the anthology is to place me on their subscription list, using the address posted in the news and announcements forum.

It has been a pleasure to work with Elizabeth Kolbert, who last year won a prestigious award from the National Academy of Sciences for her three-part *New Yorker* article "The Climate of Man." I'm very grateful to Amanda Cook and Meagan Stacey at Houghton Mifflin Harcourt for their behind-the-scenes work on this collection. And, as always, I'm more indebted than I can possibly say to my awesome and beauteous wife, Anne Nolan.

TIM FOLGER

Introduction

A FEW YEARS AGO I found myself standing on top of the Greenland ice sheet, holding a rubber hose. I had to come to Greenland to visit a bare-bones research station called Swiss Camp, which consists of a few tents mounted on a wooden platform on the snow. Before I arrived, midway through the field season, the director of the camp, Konrad Steffen, had tried to jackhammer out the ice that had accumulated in the tents since the previous summer. He had been only partially successful, so underneath the tables, where normally you would put your legs, were frozen blocks containing bits of old equipment.

Steffen established Swiss Camp to track conditions on the ice sheet. While I was visiting, he needed to replace a GPS receiver that had fallen over about ten miles from the station. This, in turn, entailed drilling a series of holes thirty feet into the ice. The drilling was done thermally, using a steam hose. Guiding the hose requires no particular skill, which is why I ended up taking a turn, along with Steffen and his graduate students. Not far from where we were working, Alfred Wegener, the German scientist who first proposed the theory of continental drift, died while on a meteorological expedition in 1930. He was buried somewhere in the ice sheet, and a long-running riff at Swiss Camp concerned stumbling on his body. "It's Wegener," one of the students joked, as the drill worked its way downward. Completing just two holes took us six hours. On the way back to camp, the snowmobile that the students were driving caught fire.

Science is not, on the whole, a glamorous enterprise. A single set of data points routinely represents months in the lab or years out in the field. Even the most productive researchers, like Steffen, are constantly getting stuck and having to dig — or jackhammer — their way out. (Steffen built Swiss Camp nearly twenty years ago, and in that time, owing to the movement of the ice, it has drifted more than a mile.) Meanwhile, most researchers' labors end up wasted, lost on hypotheses that were ill-conceived or simply unlucky. And yet they keep at it, now and then with fatal consequences. The difficulty of the work is essential to it; the true subject of science, one could argue, is the obduracy of reality.

And what goes for science also goes for science writing, or at least for the best of it. The pieces in this volume are less about answers than about questions: why does time move only in one direction? How did life begin? What happened to the Neanderthals? The stories are exciting but also demanding. They take us places that — in some cases literally, in some metaphorically — are hard to get to. They ask us to look at the world in a new way.

Frederick Kaufman's "Wasteland" begins with an ordinary family living in what should be an ordinary suburban home. Before you finish the first paragraph, excrement is bubbling up in the back yard. Kaufman decides to find out what happens to human waste, and he pursues the matter far beyond the point of politesse. He watches as shit is transformed into sludge, then dried into "biosolids" and made into fertilizer, which, he is told, is sold in the Middle East. His chronicle is fascinating and revolting and makes it hard to flush in the same mindless fashion ever again.

The stories collected here cover a wide range of disciplines, from biology to psychology to cosmology. The method I used for choosing them is perhaps best described as unscientific. (There is, as far as I know, still no formula for great writing.) Some pieces stood out for their provocative ideas, others for the authorial voice. Others were selected because of the vividness of their characters.

Atul Gawande's haunting "The Itch" tells the story of M., a woman who obsessively scratches herself. One day M. wakes up to find a greenish liquid trickling down her face. She has scratched through her scalp, exposing her brain.

Gary Wolf's profile of the futurist Ray Kurzweil is about a man who plans to live forever. Seriously. Kurweil is a wildly successful inventor; in the 1970s he developed the first computer capable of reading different fonts, and in the 1980s he created one of the first speech-recognition programs. He now takes up to 210 vitamin pills per day and receives intravenous anti-aging treatments once a week. His goal is to survive to the "singularity," defined as the moment when computers become intelligent enough to take charge of their own destinies. At that point Kurzweil intends to download his personality and "live" on as information.

The tenacity of scientists — and the recalcitrance of their subject matter — is nowhere more apparent than at the Large Hadron Collider, which is supposed to go into operation in 2009. The LHC occupies a circular tunnel, seventeen miles in circumference, under the Swiss-French border. Some ten thousand physicists and engineers from nearly a hundred countries collaborated on its design and construction. If all goes well — there have been several well-publicized mishaps along the way — the collider will accelerate protons to 99.9999991 percent of the speed of light, then smash them together inside enormous detectors. It is the most ambitious physics experiment, indeed perhaps the most ambitious experiment of any kind, ever undertaken.

The LHC is expected to provide final confirmation of what's known as the Standard Model, a set of equations that explain the behavior of most — but, significantly, not all — known forces and forms of matter. Beyond that, it's hoped that the LHC will provide clues to what's often referred to as the "new physics" — physics that may include anything from extra dimensions to multiple universes. I visited the LHC a few years ago, when its 1,600 superconducting magnets were still being tested. Descending into the tunnel, I felt as if I were entering the underworld, where the secrets of the future are kept. What struck me most in my conversations with scientists involved in the project was how eager they were to be perplexed. What they were really hoping, they told me, was that the detectors would pick up something totally baffling — evidence that reality is even more perverse than they could have imagined. In this way the LHC will either take physics into a new realm or demonstrate that we've bumped up against the discipline's limits.

New physics (and new limits) are the subject of Adam Frank's "The Day Before Genesis." Frank examines theories that attempt to answer some of the most fundamental questions of cosmology, such as what happened before the Big Bang. The theories involve branes, multiple universes, and, most disconcerting of all, the possibility that time itself does not exist. Unfortunately, Frank notes, while the theories are tantalizing, they remain — at least for the foreseeable future — unprovable. In "Blown Apart," Keay Davidson looks at another unexplained phenomenon — dark energy — and asks whether, in the age of projects like the LHC, it makes sense to think of discoveries as belonging to a single scientist. And in "Reality Tests," Joshua Roebke visits a research institute in Vienna where physicists are probing some of the weirdest of the many weird aspects of quantum mechanics. The head of the institute advises him that it's probably a mistake even to think that it's possible to distinguish between reality and information.

A different kind of limit — the limit of human rationality — is the focus of Virginia Postrel's timely "Pop Psychology." Postrel chronicles a series of lab experiments designed to test how people make financial decisions. In the experiments, subjects are asked trade securities of known value. Time and time again, as the trading progresses, they run the price of the security up to a level far above its value until, toward the end of the experiment, the price finally crashes. "Bubbles happen," Postrel writes, "even in the most controlled conditions."

"Is Google Making Us Stupid?" by Nicholas Carr poses a question that has nagged at many of us in between Internet searches. Does the way we obtain information on the Web — in short, often disconnected bursts — affect the way we process other material and, even more fundamentally, the way we think? (If you are still reading this, perhaps you're immune to Google's effect.) "We are not only *what* we read," Maryanne Wolf, a developmental psychologist at Tufts, tells Carr. "We are *how* we read."

Just as 2009 is a significant year for the future of science, thanks to the LHC, it is also, thanks to Charles Darwin, a key year in the history of science. Two hundred years ago the naturalist was born into a wealthy family in the town of Shrewsbury, in central England. A hundred and fifty years ago he published *On the Origin of Species.* (The first printing, of 1,250 copies, was oversubscribed when it went on sale.) More than twenty years of painstaking research went into *On the Origin of Species,* which either destroyed Darwin's health — one theory holds that he was infected with a parasitic disease during his travels on the HMS *Beagle* — or, more likely, exacerbated his tendency toward psychosomatic ailments. While working on his theory of natural selection, Darwin suffered from, among other things, headaches, vision problems, vomiting, cramps, and heart palpitations, and on the day he finished the manuscript, he was, according to one of his letters, "covered with a rash & fiery Boils."

A century and a half after the publication of *On the Origin of Species*, Darwin's ideas seem ever more central to our culture, even as their implications continue to challenge us. In "Minds of Their Own," Virginia Morell visits a series of brainy animals, including Alex, the famous parrot, who demonstrates his grasp of categories like shape and color, and Betsy, a border collie who understands more than three hundred words. Morell finds that humans have no monopoly on abstract thinking. (Darwin, she notes, thought even earthworms capable of cognition.) Stephen S. Hall, in "Last of the Neanderthals," looks at hominid evolution. Did our ancestors interbreed with their stockier cousins or kill them off? In "Darwin and the Meaning of Flowers," Oliver Sacks reflects on the naturalist's extensive botanical investigations. And in "Contagious Cancer," David Quammen applies Darwin's principles to the study of disease. Much like species, cancers, it turns out, evolve, in some cases into malignancies that can spread from animal to animal. This kind of contagious cancer seems to be driving the Tasmanian devil to extinction.

In "How We Evolve," Benjamin Phelan looks at the widely held belief that human evolution ended with the invention of agriculture, some ten thousand years ago. Recent advances in genomics have enabled researchers to test this belief. The results suggest that human evolution has actually been *speeding up* since people started planting crops. Phelan quotes the paleoanthropologist John Hawks, who notes that the new data continue to be resisted, even by many biologists. "Darwinism is about natural selection," Hawks observes. "But these people don't believe in natural selection — except way back when, when chimps and humans were the same."

Most science writing is done by scientists for scientists and appears in journals with names like *Molecular Biology* and *Physical Review B*. To nonscientists, such writing can seem impossibly daunting, almost as if it were written in a different language. Because of this, a lot of important scientific discoveries never reach the general public. Meanwhile, what seems compelling to a journalist is often hopelessly simplistic to a physicist or a biologist. The consequences of this language gap are serious. In the scientific community it is now generally agreed that we are in the process of determining what the world will look like for tens of thousands of years. Yet even many well-informed nonscientists are only dimly aware of what is happening.

As recently as 1927, the world's population stood at 2 billion people. It hit 3 billion in 1959. After another fifteen years it reached 4 billion, and thirteen years later it reached 5 billion. The 6 billion mark was passed in 1999, and it is expected that the 7 billion mark will be reached in February 2012. There are many forecasts of when and at what number world population will peak; the United Nations estimates that it will reach 9 billion around the middle of the century and then begin to level off.

The growth in global population has been made possible by all sorts of scientific and technological breakthroughs — synthetic fertilizers, smallpox vaccine, refrigeration, penicillin, high-yield seeds. These (and other, related) breakthroughs have allowed people — collectively and individually — to consume ever-greater quantities of the world's resources. Nearly half of the planet's land surface has now been transformed by humans. Similarly, humans use more than half of the world's readily accessible fresh water. In our chemical plants, we fix more atmospheric nitrogen than all natural terrestrial processes combined and, through our fisheries, we remove more than a third of the primary production of temperate coastal waters. The net result of all this is that wherever you look in the world today — including deep in the oceans — you can find evidence of human impacts.

The period since the last ice age ended, about ten thousand years ago, is usually referred to as the Holocene, or "wholly recent" epoch. But many argue that this term no longer makes sense, that we have entered a new era in which the Earth is being shaped as much by people as by geology. Paul Crutzen, the Dutch chemist who received a Nobel Prize for his work on ozone depletion, has dubbed this era the Anthropocene. "Human activities have become so pervasive and profound that they rival the great forces of Nature," he has written.

In "Back to the Future," J. Madeleine Nash takes the measure of our situation by considering the distant past. She visits Wyoming's Big Horn Basin with researchers who are hunting for fossils from a period known as the Paleocene-Eocene Thermal Maximum, or PETM. During the PETM, roughly 55 million years ago, massive amounts of carbon dioxide were released into the atmosphere, from where no one is quite sure. One result was a sudden and dramatic rise in Earth's temperature — the record from the PETM shows that Wyoming acquired a climate like that now found in South Florida. Another result was an abrupt change in ocean chemistry that drove many marine organisms to extinction. (When carbon dioxide dissolves in water, it forms a weak acid.) Today, of course, humans are releasing massive amounts of carbon into the atmosphere by burning fossil fuels. The result may be a future that looks uncomfortably like the PETM.

Michael Specter, in "Big Foot," looks at what could be done to reduce carbon dioxide emissions. He visits a supermarket in London that's begun to label food items with their "carbon footprint" — it takes 75 grams of carbon to produce a bag of potato chips — and talks to the head of the Chicago Climate Exchange, a market for trading carbon credits. John Broome, in "The Ethics of Climate Change," takes up the question of what *should* be done to reduce emissions. This question, he points out, may be posed in technical terms, but answering it necessarily involves moral calculations. Is it ethical to value the well-being of future generations, which will have to live with the consequences of our actions, any less than we value our own?

The title of this book, *The Best American Science and Nature Writing,* suggests that science and nature are distinct, but the defining feature of the Anthropocene is that this distinction has broken down. David Grimm, in "The Mushroom Cloud's Silver Lining," observes that all living things now carry traces of aboveground nuclear testing, in the form of elevated levels of carbon 14. (Researchers have found a way to use this extra ^{14}C to determine whether adults grow new neurons or shed old fat.)

Patrick Symmes makes his thirteenth trip to Cuba for "Red Is the New Green." Traveling around the country — one of the very few that are developing (or, in this case, perhaps, not developing) at a sustainable rate — he loses his wallet to a pair of thieves, blows his ear out scuba diving, and spends a night waiting for sea turtles to lay their eggs on the beach. He finds neglect and repression to be the only forces protecting what's left of Cuba's once-wondrous biodiversity, and he wonders what will happen once those forces are gone.

In "To Take Wildness in Hand," Michelle Nijhuis takes up the case of the Florida torreya or, as it is more colorfully known, the stinking cedar. Fewer than a thousand of the trees survive in what was once their native habitat, and now, with those survivors endangered by global warming, some biologists want to move the tree north. But can a species threatened by human intervention be saved by yet more intervention? And if the stinking cedar can be rescued, how much good would that really do? How many other species can humans realistically expect to move around the globe?

A consistent theme in the pieces collected here — and probably in all good science writing — is the contingent nature of the world as we experience it. Life on Earth in 2009 is a product of (among other things) 3 billion years of evolution, several million years of hominid evolution, ten thousand years of agriculture, and a couple of hundred years of industrialization. At any stage along the way, if events had taken a different turn, we might not be here to consider them.

Everything about us — our minds included — is a product of these processes. But that doesn't mean we are always able to understand them. What scientists have shown us over and over again is that the world is different from what we think it is, different from what we'd *like* it to be. This is, ultimately, the reason science matters: it's the one method we've found to test what we believe against the intractability of what *is*.

The power of science is something to which we have probably grown too accustomed. We've adjusted to its extraordinary benefits, so much so that we can barely imagine our lives without them. And this makes it hard for us to recognize how exceptional our situation is. Now that we control the planet, we need to learn to exercise this control responsibly, or at least self-protectively. As amazing as modern technology is, there are still limits, which, to come full circle, are set by the natural world. To pretend otherwise is to fall back into unscientific reverie, in this case, ironically, about science itself.

The very first essay in this collection is "Faustian Economics" by

Wendell Berry, who meditates on the finite terms of our existence and the basic human impulse to reject those terms: "We have insistently, and with relief, defined ourselves as animals or as 'higher animals.' But to define ourselves as animals, given our specifically human powers and desires, is to define ourselves as *limitless* animals — which is of course a contradiction in terms." The ultimate constraint on our ambitions, he observes, is inescapable: "We are not likely to be granted another world to plunder in compensation for our pillage of this one."

I am grateful to have had the opportunity to select the pieces for this book. I'd like to thank the series editor, Tim Folger, and the executive editor of Houghton Mifflin Harcourt Amanda Cook, who did the real labor in putting this volume together.

ELIZABETH KOLBERT

The Best American Science
and Nature Writing 2009

WENDELL BERRY

Faustian Economics

FROM *Harper's Magazine*

THE GENERAL REACTION to the apparent end of the era of cheap fossil fuel, as to other readily foreseeable curtailments, has been to delay any sort of reckoning. The strategies of delay so far have been a sort of willed oblivion, or visions of large profits to the manufacturers of such "biofuels" as ethanol from corn or switch-grass, or the familiar unscientific faith that "science will find an answer." The dominant response, in short, is a dogged belief that what we call the American Way of Life will prove somehow indestructible. We will keep on consuming, spending, wasting, and driving, as before, at any cost to anything and everybody but ourselves.

This belief was always indefensible — the real names of global warming are Waste and Greed — and by now it is manifestly foolish. But foolishness on this scale looks disturbingly like a sort of national insanity. We seem to have come to a collective delusion of grandeur, insisting that all of us are "free" to be as conspicuously greedy and wasteful as the most corrupt of kings and queens. (Perhaps by devoting more and more of our already abused cropland to fuel production we will at last cure ourselves of obesity and become fashionably skeletal, hungry but — thank God! — still driving.)

The problem with us is not only prodigal extravagance but also an assumed limitlessness. We have obscured the issue by refusing to see that limitlessness is a godly trait. We have insistently, and with relief, defined ourselves as animals or as "higher animals." But to define ourselves as animals, given our specifically human powers and desires, is to define ourselves as *limitless* animals — which of course is a contradiction in terms. Any definition is a limit, which is why the God of Exodus refuses to define Himself: "I am that I am."

Even so, that we have founded our present society upon delusional assumptions of limitlessness is easy enough to demonstrate. A recent "summit" in Louisville, Kentucky, was entitled "Unbridled Energy: The Industrialization of Kentucky's Energy Resources." Its subjects were "clean-coal generation, biofuels, and other cutting-edge applications," the conversion of coal to "liquid fuels," and the likelihood that all this will be "environmentally friendly." These hopes, which "can create jobs and boost the nation's security," are to be supported by government "loan guarantees . . . investment tax credits and other tax breaks." Such talk we recognize as completely conventional. It is, in fact, a tissue of clichés that is now the common tongue of promoters, politicians, and journalists. This language does not allow for any computation or speculation as to the *net* good of anything proposed. The entire contraption of "Unbridled Energy" is supported only by a rote optimism: "The United States has 250 billion tons of recoverable coal reserves — enough to last 100 years even at double the current rate of consumption." We humans have inhabited the earth for many thousands of years, and now we can look forward to surviving for another hundred by doubling our consumption of coal? This is national security? The world-ending fire of industrial fundamentalism may already be burning in our furnaces and engines, but if it will burn for a hundred more years, that will be fine. Surely it would be better to intend straightforwardly to contain the fire and eventually put it out! But once greed has been made an honorable motive, then you have an economy without limits. It has no place for temperance or thrift or the ecological law of return. It will do anything. It is monstrous by definition.

In keeping with our unrestrained consumptiveness, the commonly accepted basis of our economy is the supposed possibility of limitless growth, limitless wants, limitless wealth, limitless natural resources, limitless energy, and limitless debt. The idea of a limitless economy implies and requires a doctrine of general human limitlessness: *all* are entitled to pursue without limit whatever they conceive as desirable — a license that classifies the most exalted Christian capitalist with the lowliest pornographer.

This fantasy of limitlessness perhaps arose from the coincidence of the Industrial Revolution with the suddenly exploitable resources of the New World — though how the supposed limitlessness of resources can be reconciled with their exhaustion is not clear. Or perhaps it comes from the contrary apprehension of the world's "smallness," made possible by modern astronomy and high-speed transportation. Fear of the smallness of our world and its life may lead to a kind of claustrophobia and thence, with apparent reasonableness, to a desire for the "freedom" of limitlessness. But this desire, paradoxically, reduces everything. The life of this world is small to those who think it is, and the desire to enlarge it makes it smaller, and can reduce it finally to nothing.

However it came about, this credo of limitlessness clearly implies a principled wish not only for limitless possessions but also for limitless knowledge, limitless science, limitless technology, and limitless progress. And, necessarily, it must lead to limitless violence, waste, war, and destruction. That it should finally produce a crowning cult of political limitlessness is only a matter of mad logic.

The normalization of the doctrine of limitlessness has produced a sort of moral minimalism: the desire to be efficient at any cost, to be unencumbered by complexity. The minimization of neighborliness, respect, reverence, responsibility, accountability, and self-subordination — this is the culture of which our present leaders and heroes are the spoiled children.

Our national faith so far has been: "There's always more." Our true religion is a sort of autistic industrialism. People of intelligence and ability seem now to be genuinely embarrassed by any solution to any problem that does not involve high technology, a great expenditure of energy, or a big machine. Thus an X marked on a paper ballot no longer fulfills our idea of voting. One problem with this state of affairs is that the work now most needing to be done — that of neighborliness and caretaking — cannot be done by remote control with the greatest power on the largest scale. A second problem is that the economic fantasy of limitlessness in a limited world calls fearfully into question the value of our monetary wealth, which does not reliably stand for the real wealth of land, resources, and workmanship but instead wastes and depletes it.

That human limitlessness is a fantasy means, obviously, that its

life expectancy is limited. There is now a growing perception, and
not just among a few experts, that we are entering a time of inescapable limits. We are not likely to be granted another world to
plunder in compensation for our pillage of this one. Nor are we
likely to believe much longer in our ability to outsmart, by means of
science and technology, our economic stupidity. The hope that we
can cure the ills of industrialism by the homeopathy of more technology seems at last to be losing status. We are, in short, coming under pressure to understand ourselves as limited creatures in a limited world.

This constraint, however, is not the condemnation it may seem.
On the contrary, it returns us to our real condition and to our human heritage, from which our self-definition as limitless animals
has for too long cut us off. Every cultural and religious tradition
that I know about, while fully acknowledging our animal nature,
defines us specifically as *humans* — that is, as animals (if the word
still applies) capable of living not only within natural limits but
also within cultural limits, self-imposed. As earthly creatures, we
live, because we must, within natural limits, which we may describe
by such names as "earth" or "ecosystem" or "watershed" or "place."
But as humans, we may elect to respond to this necessary place-
ment by the self-restraints implied in neighborliness, stewardship,
thrift, temperance, generosity, care, kindness, friendship, loyalty,
and love.

In our limitless selfishness, we have tried to define "freedom,"
for example, as an escape from all restraint. But, as my friend Bert
Hornback has explained in his book *The Wisdom in Words,* "free" is
etymologically related to "friend." These words come from the
same Indo-European root, which carries the sense of "dear" or "beloved." We set our friends free by our love for them, with the implied restraints of faithfulness or loyalty. And this suggests that our
"identity" is located not in the impulse of selfhood but in deliberately maintained connections.

Thinking of our predicament has sent me back again to Christo-
pher Marlowe's *Tragical History of Doctor Faustus.* This is a play of the
Renaissance; Faustus, a man of learning, longs to possess "all Nature's treasury," to "Ransack the ocean . . . / And search all corners
of the new-found world." To assuage his thirst for knowledge and

power, he deeds his soul to Lucifer, receiving in compensation for twenty-four years the services of the subdevil Mephistophilis, nominally Faustus's slave but in fact his master. Having the subject of limitlessness in mind, I was astonished on this reading to come upon Mephistophilis's description of hell. When Faustus asks, "How comes it then that thou art out of hell?" Mephistophilis replies, "Why, this is hell, nor am I out of it." And a few pages later he explains:

> Hell hath no limits, nor is circumscribed
> In one self place, but where we [the damned] are is hell,
> And where hell is must we ever be.

For those who reject heaven, hell is everywhere, and thus is limitless. For them, even the thought of heaven is hell.

It is only appropriate, then, that Mephistophilis rejects any conventional limit: "Tut, Faustus, marriage is but a ceremonial toy. If thou lovest me, think no more of it." Continuing this theme, for Faustus's pleasure the devils present a sort of pageant of the seven deadly sins, three of which — Pride, Wrath, and Gluttony — describe themselves as orphans, disdaining the restraints of parental or filial love.

Seventy or so years later, and with the issue of the human definition more than ever in doubt, John Milton in Book VII of *Paradise Lost* returns again to a consideration of our urge to know. To Adam's request to be told the story of creation, the "affable Archangel" Raphael agrees "to answer thy desire/Of knowledge *within bounds* [my emphasis]," explaining that

> Knowledge is as food, and needs no less
> Her temperance over appetite, to know
> In measure what the mind may well contain;
> Oppresses else with surfeit, and soon turns
> Wisdom to folly, as nourishment to wind.

Raphael is saying, with angelic circumlocution, that knowledge without wisdom, limitless knowledge, is not worth a fart; he is not a humorless archangel. But he also is saying that knowledge without measure, knowledge that the human mind cannot appropriately use, is mortally dangerous.

I am well aware of what I risk in bringing this language of religion into what is normally a scientific discussion. I do so because I doubt that we can define our present problems adequately, let alone solve them, without some recourse to our cultural heritage. We are, after all, trying now to deal with the failure of scientists, technicians, and politicians to "think up" a version of human continuance that is economically probable and ecologically responsible, or perhaps even imaginable. If we go back into our tradition, we are going to find a concern with religion, which at a minimum shatters the selfish context of the individual life, and thus forces a consideration of what human beings are and ought to be.

This concern persists at least as late as our Declaration of Independence, which holds as "self-evident, that all men are created equal; that they are endowed by their Creator with certain unalienable rights." Thus among our political roots we have still our old preoccupation with our definition as humans, which in the Declaration is wisely assigned to our Creator; our rights and the rights of all humans are not granted by any human government but are innate, belonging to us by birth. This insistence comes not from the fear of death or even extinction but from the ancient fear that in order to survive we might become inhuman or monstrous.

And so our cultural tradition is in large part the record of our continuing effort to understand ourselves as beings specifically human: to say that, as humans, we must do certain things and we must not do certain things. We must have limits or we will cease to exist as humans; perhaps we will cease to exist, period. At times, for example, some of us humans have thought that human beings, properly so called, did not make war against civilian populations, or hold prisoners without a fair trial, or use torture for any reason.

Some of us would-be humans have thought too that we should not be free at anybody else's expense. And yet in the phrase "free market," the word "free" has come to mean unlimited economic power for some, with the necessary consequence of economic powerlessness for others. Several years ago, after I had spoken at a meeting, two earnest and obviously troubled young veterinarians approached me with a question: How could they practice veterinary medicine without serious economic damage to the farmers who were their clients? Underlying their question was the fact that for a long time veterinary help for a sheep or a pig has been likely to cost more than the animal is worth. I had to answer that, in my opinion, so long as their practice relied heavily on selling patented drugs, they had no choice, since the market for medicinal drugs was entirely controlled by the drug companies, whereas most farmers had no control at all over the market for agricultural products. My questioners were asking in effect if a predatory economy can have a beneficent result. The answer too often is No. And that is because there is an absolute discontinuity between the economy of the seller of medicines and the economy of the buyer, as there is in the health industry as a whole. The drug industry is interested in the survival of patients, we have to suppose, because surviving patients will continue to consume drugs.

Now let us consider a contrary example. Recently, at another meeting, I talked for some time with an elderly, and some would say an old-fashioned, farmer from Nebraska. Unable to farm any longer himself, he had rented his land to a younger farmer on the basis of what he called "crop share" instead of a price paid or owed in advance. Thus, as the old farmer said of his renter, "If he has a good year, I have a good year. If he has a bad year, I have a bad one." This is what I would call community economics. It is a sharing of fate. It assures an economic continuity and a common interest between the two partners to the trade. This is as far as possible from the economy in which the young veterinarians were caught, in which the powerful are limitlessly "free" to trade, to the disadvantage, and ultimately the ruin, of the powerless.

It is this economy of community destruction that, wittingly or unwittingly, most scientists and technicians have served for the past two hundred years. These scientists and technicians have justified themselves by the proposition that they are the vanguard of progress, enlarging human knowledge and power, and thus they have romanticized both themselves and the predatory enterprises that they have served.

As a consequence, our great need now is for sciences and technologies of limits, of domesticity, of what Wes Jackson of the Land Institute in Salina, Kansas, has called "homecoming." These would be specifically human sciences and technologies, working, as the best humans always have worked, within self-imposed limits. The limits would be the accepted contexts of places, communities, and neighborhoods, both natural and human.

I know that the idea of such limitations will horrify some people, maybe most people, for we have long encouraged ourselves to feel at home on "the cutting edges" of knowledge and power or on some "frontier" of human experience. But I know too that we are talking now in the presence of much evidence that improvement by outward expansion may no longer be a good idea, if it ever was. It was not a good idea for the farmers who "leveraged" secure acreage to buy more during the 1970s. It has proved tragically to be a bad idea in a number of recent wars. If it is a good idea in the form of corporate gigantism, then we must ask, For whom? Faustus, who wants all knowledge and all the world for himself, is a man supremely lonely and finally doomed. I don't think Marlowe was kidding. I don't think Satan is kidding when he says in *Paradise Lost*, "Myself am Hell."

If the idea of appropriate limitation seems unacceptable to us, that may be because, like Marlowe's Faustus and Milton's Satan, we confuse limits with confinement. But that, as I think Marlowe and Milton and others were trying to tell us, is a great and potentially a fatal mistake. Satan's fault, as Milton understood it and perhaps with some sympathy, was precisely that he could not tolerate his proper limitation; he could not subordinate himself to anything whatever. Faustus's error was his unwillingness to remain "Faustus, and a man." In our age of the world it is not rare to find writers, critics, and teachers of literature, as well as scientists and technicians, who regard Satan's and Faustus's defiance as salutary and heroic.

On the contrary, our human and earthly limits, properly understood, are not confinements but rather inducements to formal elaboration and elegance, to *fullness* of relationship and meaning. Perhaps our most serious cultural loss in recent centuries is the knowledge that some things, though limited, are inexhaustible. For example, an ecosystem, even that of a working forest or farm, so long as it remains ecologically intact, is inexhaustible. A small place, as I know from my own experience, can provide opportunities of work and learning, and a fund of beauty, solace, and pleasure — in addition to its difficulties — that cannot be exhausted in a lifetime or in generations.

To recover from our disease of limitlessness, we will have to give up the idea that we have a right to be godlike animals, that we are po-

tentially omniscient and omnipotent, ready to discover "the secret
of the universe." We will have to start over, with a different and
much older premise: the naturalness and, for creatures of limited
intelligence, the necessity, of limits. We must learn again to ask how
we can make the most of what we are, what we have, what we have
been given. If we always have a theoretically better substitute available from somebody or someplace else, we will never make the
most of anything. It is hard to make the most of one life. If we each
had two lives, we would not make much of either. Or as one of my
best teachers said of people in general: "They'll never be worth a
damn as long as they've got two choices."

To deal with the problems, which after all are inescapable, of living with limited intelligence in a limited world, I suggest that we
may have to remove some of the emphasis we have lately placed on
science and technology and have a new look at the arts. For an art
does not propose to enlarge itself by limitless extension but rather
to enrich itself within bounds that are accepted prior to the work.

It is the artists, not the scientists, who have dealt unremittingly with the problem of limits. A painting, however large, must
finally be bounded by a frame or a wall. A composer or playwright
must reckon, at a minimum, with the capacity of an audience to sit
still and pay attention. A story, once begun, must end somewhere
within the limits of the writer's and the reader's memory. And of
course the arts characteristically impose limits that are artificial:
the five acts of a play, or the fourteen lines of a sonnet. Within
these limits artists achieve elaborations of pattern, of sustaining relationships of parts with one another and with the whole, that may
be astonishingly complex. And probably most of us can name a
painting, a piece of music, a poem or play or story that still grows in
meaning and remains fresh after many years of familiarity.

We know by now that a natural ecosystem survives by the same
sort of formal intricacy, ever-changing, inexhaustible, and no doubt
finally unknowable. We know further that if we want to make our
economic landscapes sustainably and abundantly productive, we
must do so by maintaining in them a living formal complexity
something like that of natural ecosystems. We can do this only by
raising to the highest level our mastery of the arts of agriculture,
animal husbandry, forestry, and, ultimately, the art of living.

It is true that insofar as scientific experiments must be con-

ducted within carefully observed limits, scientists also are artists.
But in science one experiment, whether it succeeds or fails, is logi-
cally followed by another in a theoretically infinite progression. According to the underlying myth of modern science, this progression is always replacing the smaller knowledge of the past with the
larger knowledge of the present, which will be replaced by the yet
larger knowledge of the future.

In the arts, by contrast, no limitless sequence of works is ever implied or looked for. No work of art is necessarily followed by a second work that is necessarily better. Given the methodologies of
science, the law of gravity and the genome were bound to be discovered by somebody; the identity of the discoverer is incidental to
the fact. But it appears that in the arts there are no second chances.
We must assume that we had one chance each for *The Divine Comedy*
and *King Lear.* If Dante and Shakespeare had died before they
wrote those poems, nobody ever would have written them.

The same is true of our arts of land use, our economic arts, which
are our arts of living. With these it is once-for-all. We will have
no chance to redo our experiments with bad agriculture leading
to soil loss. The Appalachian mountains and forests we have destroyed for coal are gone forever. It is now and forevermore too late
to use thriftily the first half of the world's supply of petroleum. In
the art of living we can only start again with what remains.

And so, in confronting the phenomenon of "peak oil," we are really confronting the end of our customary delusion of "more."
Whichever way we turn, from now on, we are going to find a limit
beyond which there will be no more. To hit these limits at top
speed is not a rational choice. To start slowing down, with the idea
of avoiding catastrophe, is a rational choice, and a viable one if we
can recover the necessary political sanity. Of course it makes sense
to consider alternative energy sources, provided *they* make sense.
But also we will have to re-examine the economic structures of our
lives, and conform them to the tolerances and limits of our earthly
places. Where there is no more, our one choice is to make the most
and the best of what we have.

JOHN BROOME

The Ethics of Climate Change

FROM *Scientific American*

WHAT SHOULD WE DO about climate change? The question is an ethical one. Science, including the science of economics, can help discover the causes and effects of climate change. It can also help work out what we can do about climate change. But what we *should* do is an ethical question.

Not all "should" questions are ethical. "How should you hold a golf club?" is not, for instance. The climate question is ethical, however, because any thoughtful answer must weigh conflicting interests among different people. If the world is to do something about climate change, some people — chiefly the better-off among the current generation — will have to reduce their emissions of greenhouse gases to save future generations from the possibility of a bleak existence in a hotter world. When interests conflict, "should" questions are always ethical.

Climate change raises a number of ethical questions. How should we — all of us living today — evaluate the well-being of future generations, given that they are likely to have more material goods than we do? Many people, some living, others yet to be born, will die from the effects of climate change. Is each death equally bad? How bad are those deaths collectively? Many people will die before they bear children, so climate change will prevent the existence of children who would otherwise have been born. Is their nonexistence a bad thing? By emitting greenhouse gases, are the rich perpetrating an injustice on the world's poor? How should we respond to the small but real chance that climate change could lead to worldwide catastrophe?

Many ethical questions can be settled by common sense. Sophisticated philosophy is rarely needed. All of us are to some extent equipped to face up to the ethical questions raised by climate change. For example, almost everyone recognizes (with some exceptions) the elementary moral principle that you should not do something for your own benefit if it harms another person. True, sometimes you cannot avoid harming someone, and sometimes you may do it accidentally without realizing it. But whenever you cause harm, you should normally compensate the victim.

Climate change will cause harm. Heat waves, storms, and floods will kill many people and harm many others. Tropical diseases, which will increase their range as the climate warms, will exact their toll in human lives. Changing patterns of rainfall will lead to local shortages of food and safe drinking water. Large-scale human migrations in response to rising sea levels and other climate-induced stresses will impoverish many people. As yet, few experts have predicted specific numbers, but some statistics suggest the scale of the harm that climate change will cause. The European heat wave of 2003 is estimated to have killed 35,000 people. In 1998 floods in China adversely affected 240 million. The World Health Organization estimates that as long ago as 2000 the annual death toll from climate change had already reached more than 150,000.

In going about our daily lives, each of us causes greenhouse gases to be emitted. Driving a car, using electric power, buying anything whose manufacture or transport consumes energy — all those activities generate greenhouse gases that contribute to climate change. In that way, what we each do for our own benefit harms others. Perhaps at the moment we cannot help it, and in the past we did not realize we were doing it. But the elementary moral principle I mentioned tells us we should try to stop doing it and compensate the people we harm.

This same principle also tells us that what we should do about climate change is not just a matter of weighing benefits against costs — although it is partly that. Suppose you calculate that the benefit to you and your friends of partying until dawn exceeds the harm done to your neighbor by keeping her awake all night. It does not follow that you should hold your party. Similarly, think of an industrial project that brings benefits in the near future but emits greenhouse gases that will harm people decades hence. Again suppose the benefits exceed the costs. It does not follow that the project should go ahead; indeed, it may be morally wrong. Those who benefit from it should not impose its costs on others who do not.

Ethics of Costs and Benefits

But even if weighing costs against benefits does not entirely answer the question of what should be done about climate change, it is an essential part of the answer. The costs of mitigating climate change are the sacrifices the present generation will have to make to reduce greenhouse gases. We will have to travel less and better insulate our homes. We will have to eat less meat. We will have to live less lavishly. The benefits are the better lives that future people will lead: they will not suffer so much from the spread of deserts, from the loss of their homes to the rising sea, or from floods, famines, and the general impoverishment of nature.

Weighing benefits to some people against costs to others is an ethical matter. But many of the costs and benefits of mitigating climate change present themselves in economic terms, and economics has useful methods of weighing benefits against costs in complex cases. So here economics can work in the service of ethics.

The ethical basis of cost-benefit economics was recognized recently in a major report, the *Stern Review on the Economics of Climate Change*, by Nicholas Stern and his colleagues at the UK Treasury. The *Stern Review* concentrates mainly on comparing costs and benefits, and it concludes that the benefit that would be gained by reducing emissions of greenhouse gases would be far greater than the cost of reducing them. Stern's work has provoked a strong reaction from economists for two reasons. First, some economists think economic conclusions should not be based on ethical premises. Second, the review favors strong and immediate action to control emissions, whereas other economic studies, such as one by William Nordhaus of Yale University, have concluded that the need to act is not so urgent.

Those two issues are connected. Stern's conclusion differs from Nordhaus's principally because, on ethical grounds, Stern uses a lower "discount rate." Economists generally value future goods less than present ones: they discount future goods. Furthermore, the more distant the future in which goods become available, the more the goods are discounted. The discount rate measures how fast the value of goods diminishes with time. Nordhaus discounts at roughly 6 percent a year; Stern discounts at 1.4 percent. The effect is that Stern gives a present value of $247 billion for having, say, a trillion dollars' worth of goods a century from now. Nordhaus values having those same goods in 2108 at just $2.5 billion today. Thus, Stern attaches nearly one hundred times as much value as Nordhaus does to having any given level of costs and benefits one hundred years from now.

The difference between the two economists' discount rates is enough to explain the difference between their conclusions. Most of the costs of controlling climate change must be borne in the near future, when the present generation must sacrifice some of its consumption. The benefits will mostly come a century or two from now. Because Stern judges the present value of those benefits to be higher than Nordhaus does, Stern can justify spending more today on mitigating climate change than Nordhaus can.

The Richer Future

Why discount future goods at all? The goods in question are the material goods and services that people consume — bicycles, food, banking services, and so on. In most of the scenarios predicted for climate change, the world economy will continue to grow. Hence, future people will on average possess more goods than present people do. The more goods you already have, the less valuable are further goods, and so it is sound economic logic to discount them. To have one bathroom in your house is a huge improvement to your life; a second bathroom is nice but not so life-changing. Goods have "diminishing marginal value," as economists put it.

But there may be a second, purely ethical reason for discounting goods that come to relatively rich people. According to an ethical theory known as prioritarianism, a benefit — by which I mean an increase in an individual's well-being — that comes to a rich person should be assigned less social value than the same benefit would have if it had come to a poor person. Prioritarianism gives priority to the less well off. According to an alternative ethical theory known as utilitarianism, however, a benefit has the same value no matter who receives it. Society should simply aim to maximize the total of people's well-being, no matter how that total is distributed across the population.

What should the discount rate be? What determines how *fast* the value of having goods in the future diminishes as the future time in question becomes more remote? That depends, first, on some nonethical factors. Among them is the economy's rate of growth, which measures how much better off, on average, people will be in the future than they are today. Consequently, it determines how much less benefit future people will derive from additional material goods than people would derive now from those same goods. A fast growth rate makes for a high discount rate.

The discount rate also depends on an ethical factor. How should benefits to those future, richer people be valued in comparison to our own? If prioritarianism is right, the value attached to future people's benefits should be less than the value of our benefits, because future people will be better off than we are. If utilitarianism is right, future people's benefits should be valued equally with ours. Prioritarianism therefore makes for a relatively high discount rate; utilitarianism makes for a lower one.

The debate between prioritarians and utilitarians takes a curious, even poignant turn in this context. Most debates about inequality take place among the relatively rich, when they consider what sacrifices they should make for the relatively poor. But when we think about future people, we are considering what sacrifices we, the relatively poor, should make for the later relatively rich. Usually prioritarianism demands more of the developed countries than utilitarianism does. In this case, it demands less.

Temporal Distance

Another ethical consideration also affects the discount rate. Some philosophers think we should care more about people who live close to us in time than about those who live in the more distant future, just because of their temporal distance from us. If those philosophers are right, future well-being should be discounted just because it comes in the future. This position is called pure discounting. It implies we should give less importance to the death of a ten-year-old one hundred years in the future than to the death of a ten-year-old now. An opposing view is that we should be temporally impartial, insisting that the mere date on which a harm occurs makes no difference to its value. Pure discounting makes for a relatively high discount rate; temporal impartiality makes for a lower one.

To determine the right discount rate, therefore, the economist must answer at least two ethical questions. Which should we accept: prioritarianism or utilitarianism? And should we adopt pure discounting or be temporally impartial?

These questions are not matters of elementary morality; they raise difficult issues in moral philosophy. Moral philosophers approach such questions by combining tight analytical argument with sensitivity to ethical intuitions. Arguments in moral philosophy are rarely conclusive, partly because we each have mutually inconsistent intuitions. All I can do as a philosopher is judge the truth as well as I can and present my best arguments in support of my judgments. Space prevents me from setting forth my arguments here, but I have concluded that prioritarianism is mistaken and that we should be temporally impartial. For more detail, see chapter 10 of my book *Weighing Goods* (1991) and section 4.3 of my book *Weighing Lives* (2004).

Market Discount Rates?

Stern reaches those same ethical conclusions. Since both tend toward low discounting, they — together with Stern's economic modeling — lead him to his 1.4 percent rate. His practical conclusion follows: the world urgently needs to take strong measures to control climate change.

Economists who oppose Stern do not deny that his practical conclusion follows from his ethical stance. They object to his ethical stance. Yet most of them decline to take any ethical position of their own, even though they favor an interest rate higher than Stern's. As I have explained, the correct discount rate depends on ethical considerations. So how can economists justify a discount rate without taking an ethical position?

They do so by taking their higher discount rate from the money market, where people exchange future money for present money, and vice versa. They adopt the money-market interest rate as their interest rate. How can that be justified?

First, some values are determined by people's tastes, which markets do reveal. The relative value of apples and oranges is determined by the tastes revealed in the fruit market. But the value that should be attached to the well-being of future generations is not determined by tastes. It is a matter of ethical judgment.

So does the money market reveal people's ethical judgments about the value of future well-being? I doubt it. The evidence shows that, when people borrow and lend, they often give less weight to their own future well-being than to their present well-being. Most of us are probably not so foolish as to judge that our own well-being is somehow less valuable in old age than in youth. Instead our behavior simply reflects our impatience to enjoy a present benefit, overwhelming whatever judgment we might make about the value of our own future. Inevitably, impatience will also overwhelm whatever high-minded arguments we might make in favor of the well-being of future generations.

But for the sake of argument, suppose people's market behavior genuinely reflected their judgments of value. How could economists then justify proclaiming an ethically neutral stance and taking the discount rate from the market? They do so, purportedly, on democratic grounds — leaving ethical judgments to the public rather than making them for themselves. The economists who criticize Stern claim the democratic high ground and accuse him of arrogantly trying to impose his own ethical beliefs on others.

They misunderstand democracy. Democracy requires debate and deliberation as well as voting. Economists — even Stern — cannot impose their beliefs on anyone. They can only make recommendations and argue for them. Determining the correct discount rate requires sophisticated theory, and we members of the public cannot do it without advice from experts. The role of economists in the democratic process is to work out that theory. They should offer their best recommendations, supported by their best arguments. They should be willing to engage in debate with one another about the ethical bases of their conclusions. Then we members of the public must reach our own decisions with the experts' help. Without their help, our choices will be uninformed and almost worthless.

Once we have made our decisions through the democratic process, society can act. That is not the job of economists. Their recommendations are inputs to the process, not the output of it. The true arrogance is imagining that you are the final arbiter of the democratic process.

Ethical considerations cannot be avoided in determining the discount rate. Climate change raises many other ethical issues, too; one crucial one, the problem of catastrophic outcomes, will require serious work in ethics to decide what sacrifices we should make to moderate climate change. Like the science of climate change, the ethics of climate change is hard. So far it leaves much to be resolved. We face ethical as well as scientific problems, and we must work to solve them.

NICHOLAS CARR

Is Google Making Us Stupid?

FROM *The Atlantic Monthly*

"DAVE, STOP. Stop, will you? Stop, Dave. Will you stop, Dave?" So the supercomputer HAL pleads with the implacable astronaut Dave Bowman in a famous and weirdly poignant scene toward the end of Stanley Kubrick's *2001: A Space Odyssey*. Bowman, having nearly been sent to a deep-space death by the malfunctioning machine, is calmly, coldly disconnecting the memory circuits that control its artificial brain. "Dave, my mind is going," HAL says, forlornly. "I can feel it. I can feel it."

I can feel it, too. Over the past few years I've had an uncomfortable sense that someone, or something, has been tinkering with my brain, remapping the neural circuitry, reprogramming the memory. My mind isn't going — so far as I can tell — but it's changing. I'm not thinking the way I used to think. I can feel it most strongly when I'm reading. Immersing myself in a book or a lengthy article used to be easy. My mind would get caught up in the narrative or the turns of the argument, and I'd spend hours strolling through long stretches of prose. That's rarely the case anymore. Now my concentration often starts to drift after two or three pages. I get fidgety, lose the thread, begin looking for something else to do. I feel as if I'm always dragging my wayward brain back to the text. The deep reading that used to come naturally has become a struggle.

I think I know what's going on. For more than a decade now, I've been spending a lot of time online, searching and surfing and sometimes adding to the great databases of the Internet. The Web has been a godsend to me as a writer. Research that once required days in the stacks or periodical rooms of libraries can now be done in minutes. A few Google searches, some quick clicks on hyperlinks, and I've got the telltale fact or pithy quote I was after. Even when I'm not working, I'm as likely as not to be foraging in the Web's info-thickets — reading and writing e-mails, scanning headlines and blog posts, watching videos and listening to podcasts, or just tripping from link to link to link. (Unlike footnotes, to which they're sometimes likened, hyperlinks don't merely point to related works; they propel you toward them.)

For me, as for others, the Net is becoming a universal medium, the conduit for most of the information that flows through my eyes and ears and into my mind. The advantages of having immediate access to such an incredibly rich store of information are many, and they've been widely described and duly applauded. "The perfect recall of silicon memory," *Wired*'s Clive Thompson has written, "can be an enormous boon to thinking." But that boon comes at a price. As the media theorist Marshall McLuhan pointed out in the 1960s, media are not just passive channels of information. They supply the stuff of thought, but they also shape the process of thought. And what the Net seems to be doing is chipping away my capacity for concentration and contemplation. My mind now expects to take in information the way the Net distributes it: in a swiftly moving stream of particles. Once I was a scuba diver in the sea of words. Now I zip along the surface like a guy on a Jet Ski.

I'm not the only one. When I mention my troubles with reading to friends and acquaintances — literary types, most of them — many say they're having similar experiences. The more they use the Web, the more they have to fight to stay focused on long pieces of writing. Some of the bloggers I follow have also begun mentioning the phenomenon. Scott Karp, who writes a blog about online media, recently confessed that he has stopped reading books altogether. "I was a lit major in college, and used to be [a] voracious book reader," he wrote. "What happened?" He speculates on the answer: "What if I do all my reading on the web not so much because the way I read has changed, i.e. I'm just seeking convenience, but because the way I THINK has changed?"

Bruce Friedman, who blogs regularly about the use of computers in medicine, also has described how the Internet has altered his mental habits. "I now have almost totally lost the ability to read and absorb a longish article on the web or in print," he wrote earlier this year. A pathologist who has long been on the faculty of the University of Michigan Medical School, Friedman elaborated on his comment in a telephone conversation with me. His thinking, he said, has taken on a "staccato" quality, reflecting the way he quickly scans short passages of text from many sources online. "I can't read *War and Peace* anymore," he admitted. "I've lost the ability to do that. Even a blog post of more than three or four paragraphs is too much to absorb. I skim it."

Anecdotes alone don't prove much. And we still await the long-term neurological and psychological experiments that will provide a definitive picture of how Internet use affects cognition. But a recently published study of online research habits, conducted by scholars from University College London, suggests that we may well be in the midst of a sea change in the way we read and think. As part of the five-year research program, the scholars examined computer logs documenting the behavior of visitors to two popular research sites, one operated by the British Library and one by a UK educational consortium, that provide access to journal articles, e-books, and other sources of written information. They found that people using the sites exhibited "a form of skimming activity," hopping from one source to another and rarely returning to any source they'd already visited. They typically read no more than one or two pages of an article or book before they would "bounce" out to another site. Sometimes they'd save a long article, but there's no evidence that they ever went back and actually read it. The authors of the study report:

> It is clear that users are not reading online in the traditional sense; indeed there are signs that new forms of "reading" are emerging as users "power browse" horizontally through titles, contents pages and abstracts going for quick wins. It almost seems that they go online to avoid reading in the traditional sense.

Thanks to the ubiquity of text on the Internet, not to mention the popularity of text-messaging on cell phones, we may well be reading more today than we did in the 1970s or 1980s, when television was our medium of choice. But it's a different kind of reading, and behind it lies a different kind of thinking—perhaps even a new sense of the self. "We are not only *what* we read," says

Maryanne Wolf, a developmental psychologist at Tufts University and the author of *Proust and the Squid: The Story and Science of the Reading Brain*. "We are *how* we read." Wolf worries that the style of reading promoted by the Net, a style that puts "efficiency" and "immediacy" above all else, may be weakening our capacity for the kind of deep reading that emerged when an earlier technology, the printing press, made long and complex works of prose commonplace. When we read online, she says, we tend to become "mere decoders of information." Our ability to interpret text, to make the rich mental connections that form when we read deeply and without distraction, remains largely disengaged.

Reading, explains Wolf, is not an instinctive skill for human beings. It's not etched into our genes the way speech is. We have to teach our minds how to translate the symbolic characters we see into the language we understand. And the media or other technologies we use in learning and practicing the craft of reading play an important part in shaping the neural circuits inside our brains. Experiments demonstrate that readers of ideograms, such as the Chinese, develop a mental circuitry for reading that is very different from the circuitry found in those of us whose written language employs an alphabet. The variations extend across many regions of the brain, including those that govern such essential cognitive functions as memory and the interpretation of visual and auditory stimuli. We can expect as well that the circuits woven by our use of the Net will be different from those woven by our reading of books and other printed works.

Sometime in 1882, Friedrich Nietzsche bought a typewriter — a Malling-Hansen Writing Ball, to be precise. His vision was failing, and keeping his eyes focused on a page had become exhausting and painful, often bringing on crushing headaches. He had been forced to curtail his writing, and he feared that he would soon have to give it up. The typewriter rescued him, at least for a time. Once he had mastered touch-typing, he was able to write with his eyes closed, using only the tips of his fingers. Words could once again flow from his mind to the page.

But the machine had a subtler effect on his work. One of Nietzsche's friends, a composer, noticed a change in the style of his writing. His already terse prose had become even tighter, more telegraphic. "Perhaps you will through this instrument even take to a new idiom," the friend wrote in a letter, noting that, in his own work, his "'thoughts' in music and language often depend on the quality of pen and paper."

"You are right," Nietzsche replied, "our writing equipment takes part in the forming of our thoughts." Under the sway of the machine, writes the German media scholar Friedrich A. Kittler, Nietzsche's prose "changed from arguments to aphorisms, from thoughts to puns, from rhetoric to telegram style."

The human brain is almost infinitely malleable. People used to think that our mental meshwork, the dense connections formed among the 100 billion or so neurons inside our skulls, was largely fixed by the time we reached adulthood. But brain researchers have discovered that that's not the case. James Olds, a professor of neuroscience who directs the Krasnow Institute for Advanced Study at George Mason University, says that even the adult mind "is very plastic." Nerve cells routinely break old connections and form new ones. "The brain," according to Olds, "has the ability to reprogram itself on the fly, altering the way it functions."

As we use what the sociologist Daniel Bell has called our "intellectual technologies" — the tools that extend our mental rather than our physical capacities — we inevitably begin to take on the qualities of those technologies. The mechanical clock, which came into common use in the fourteenth century, provides a compelling example. In *Technics and Civilization*, the historian and cultural critic Lewis Mumford described how the clock "disassociated time from human events and helped create the belief in an independent world of mathematically measurable sequences." The "abstract framework of divided time" became "the point of reference for both action and thought."

The clock's methodical ticking helped bring into being the scientific mind and the scientific man. But it also took something away. As the late MIT computer scientist Joseph Weizenbaum observed in his 1976 book, *Computer Power and Human Reason: From Judgment to Calculation,* the conception of the world that emerged from the widespread use of timekeeping instruments "remains an impoverished version of the older one, for it rests on a rejection of those direct experiences that formed the basis for, and indeed constituted, the old reality." In deciding when to eat, to work, to sleep, to rise, we stopped listening to our senses and started obeying the clock.

The process of adapting to new intellectual technologies is reflected in the changing metaphors we use to explain ourselves to ourselves. When the mechanical clock arrived, people began thinking of their brains as operating "like clockwork." Today, in the age of software, we have come to think of them as operating "like computers." But the changes, neuroscience tells us, go much deeper than metaphor. Thanks to our brain's plasticity, the adaptation occurs also at a biological level.

The Internet promises to have particularly far-reaching effects on cognition. In a paper published in 1936, the British mathematician Alan Turing proved that a digital computer, which at the time existed only as a theoretical machine, could be programmed to perform the function of any other information-processing device. And that's what we're seeing today. The Internet, an immeasurably powerful computing system, is subsuming most of our other intellectual technologies. It's becoming our map and our clock, our printing press and our typewriter, our calculator and our telephone, and our radio and TV.

When the Net absorbs a medium, that medium is recreated in the Net's image. It injects the medium's content with hyperlinks, blinking ads, and other digital gewgaws, and it surrounds the content with the content of all the other media it has absorbed. A new e-mail message, for instance, may announce its arrival as we're glancing over the latest headlines at a newspaper's site. The result is to scatter our attention and diffuse our concentration.

The Net's influence doesn't end at the edges of a computer screen, either. As people's minds become attuned to the crazy quilt of Internet media, traditional media have to adapt to the audience's new expectations. Television programs add text crawls and pop-up ads, and magazines and newspapers shorten their articles, introduce capsule summaries, and crowd their pages with easy-to-browse info-snippets. When, in March of this year, the *New York Times* decided to devote the second and third pages of every edition to article abstracts, its design director, Tom Bodkin, explained that the "shortcuts" would give harried readers a quick "taste" of the day's news, sparing them the "less efficient" method of actually turning the pages and reading the articles. Old media have little choice but to play by the new-media rules.

Never has a communications system played so many roles in our lives — or exerted such broad influence over our thoughts — as the Internet does today. Yet for all that's been written about the Net, there's been little consideration of how, exactly, it's reprogramming us. The Net's intellectual ethic remains obscure.

About the same time that Nietzsche started using his typewriter, an earnest young man named Frederick Winslow Taylor carried a stopwatch into the Midvale Steel plant in Philadelphia and began a historic series of experiments aimed at improving the efficiency of the plant's machinists. With the approval of Midvale's owners, he recruited a group of factory hands, set them to work on various metalworking machines, and recorded and timed their every movement as well as the operations of the machines. By breaking down every job into a sequence of small, discrete steps and then testing different ways of performing each one, Taylor created a set of precise instructions — an "algorithm," we might say today — for how each worker should work. Midvale's employees grumbled about the strict new regime, claiming that it turned them into little more than automatons, but the factory's productivity soared.

More than a hundred years after the invention of the steam engine, the Industrial Revolution had at last found its philosophy and its philosopher. Taylor's tight industrial choreography — his "system," as he liked to call it — was embraced by manufacturers throughout the country and, in time, around the world. Seeking maximum speed, maximum efficiency, and maximum output, factory owners used time-and-motion studies to organize their work and configure the jobs of their workers. The goal, as Taylor defined it in his celebrated 1911 treatise, *The Principles of Scientific Management,* was to identify and adopt, for every job, the "one best method" of work and thereby to effect "the gradual substitution of science for rule of thumb throughout the mechanic arts." Once his system was applied to all acts of manual labor, Taylor assured his followers, it would bring about a restructuring not only of industry but of society, creating a utopia of perfect efficiency. "In the past the man has been first," he declared; "in the future the system must be first."

Taylor's system is still very much with us; it remains the ethic of industrial manufacturing. And now, thanks to the growing power that computer engineers and software coders wield over our intellectual lives, Taylor's ethic is beginning to govern the realm of the mind as well. The Internet is a machine designed for the efficient and automated collection, transmission, and manipulation of information, and its legions of programmers are intent on finding the "one best method" — the perfect algorithm — to carry out every mental movement of what we've come to describe as "knowledge work."

Google's headquarters, in Mountain View, California — the Googleplex — is the Internet's high church, and the religion practiced inside its walls is Taylorism. Google, says its chief executive, Eric Schmidt, is "a company that's founded around the science of measurement," and it is striving to "systematize everything" it does. Drawing on the terabytes of behavioral data it collects through its search engine and other sites, it carries out thousands of experiments a day, according to the *Harvard Business Review,* and it uses the results to refine the algorithms that increasingly control how people find information and extract meaning from it. What Taylor did for the work of the hand, Google is doing for the work of the mind.

The company has declared that its mission is "to organize the world's information and make it universally accessible and useful." It seeks to develop "the perfect search engine," which it defines as something that "understands exactly what you mean and gives you back exactly what you want." In Google's view, information is a kind of commodity, a utilitarian resource that can be mined and processed with industrial efficiency. The more pieces of information we can "access" and the faster we can extract their gist, the more productive we become as thinkers.

Where does it end? Sergey Brin and Larry Page, the gifted young men who founded Google while pursuing doctoral degrees in computer science at Stanford, speak frequently of their desire to turn their search engine into an artificial intelligence, a HAL-like machine that might be connected directly to our brains. "The ultimate search engine is something as smart as people — or smarter," Page said in a speech a few years back. "For us, working on search is a way to work on artificial intelligence." In a 2004 interview with *Newsweek,* Brin said, "Certainly if you had all the world's information directly attached to your brain, or an artificial brain that was smarter than your brain, you'd be better off." Last year Page told a convention of scientists that Google is "really trying to build artificial intelligence and to do it on a large scale."

Such an ambition is a natural one, even an admirable one, for a pair of math whizzes with vast quantities of cash at their disposal and a small army of computer scientists in their employ. A fundamentally scientific enterprise, Google is motivated by a desire to use technology, in Eric Schmidt's words, "to solve problems that have never been solved before," and artificial intelligence is the hardest problem out there. Why wouldn't Brin and Page want to be the ones to crack it?

Still, their easy assumption that we'd all "be better off" if our brains were supplemented, or even replaced, by an artificial intelligence is unsettling. It suggests a belief that intelligence is the output of a mechanical process, a series of discrete steps that can be isolated, measured, and optimized. In Google's world, the world we enter when we go online, there's little place for the fuzziness of contemplation. Ambiguity is not an opening for insight but a bug to be fixed. The human brain is just an outdated computer that needs a faster processor and a bigger hard drive.

The idea that our minds should operate as high-speed data-processing machines is not only built into the workings of the Internet, it is the network's reigning business model as well. The faster we surf across the Web — the more links we click and pages we view — the more opportunities Google and other companies gain to collect information about us and to feed us advertisements. Most of the proprietors of the commercial Internet have a financial stake in collecting the crumbs of data we leave behind as we flit from link to link — the more crumbs, the better. The last thing these companies want is to encourage leisurely reading or slow, concentrated thought. It's in their economic interest to drive us to distraction.

Maybe I'm just a worrywart. Just as there's a tendency to glorify technological progress, there's a countertendency to expect the worst of every new tool or machine. In Plato's *Phaedrus,* Socrates bemoaned the development of writing. He feared that, as people came to rely on the written word as a substitute for the knowledge they used to carry inside their heads, they would, in the words of one of the dialogue's characters, "cease to exercise their memory and become forgetful." And because they would be able to "receive a quantity of information without proper instruction," they would "be thought very knowledgeable when they are for the most part quite ignorant." They would be "filled with the conceit of wisdom instead of real wisdom." Socrates wasn't wrong — the new technology did often have the effects he feared — but he was shortsighted. He couldn't foresee the many ways that writing and reading would serve to spread information, spur fresh ideas, and expand human knowledge (if not wisdom).

The arrival of Gutenberg's printing press in the fifteenth century set off another round of teeth gnashing. The Italian humanist Hieronimo Squarciafico worried that the easy availability of books would lead to intellectual laziness, making men "less studious" and weakening their minds. Others argued that cheaply printed books and broadsheets would undermine religious authority, demean the work of scholars and scribes, and spread sedition and debauchery. As the New York University professor Clay Shirky notes, "Most of the arguments made against the printing press were correct, even prescient." But, again, the doomsayers were unable to imagine the myriad blessings that the printed word would deliver.

So, yes, you should be skeptical of my skepticism. Perhaps those who dismiss critics of the Internet as Luddites or nostalgists will be proved correct, and from our hyperactive, data-stoked minds will spring a golden age of intellectual discovery and universal wisdom. Then again, the Net isn't the alphabet, and although it may replace the printing press, it produces something altogether different. The kind of deep reading that a sequence of printed pages promotes is valuable not just for the knowledge we acquire from the author's words but for the intellectual vibrations those words set off within our own minds. In the quiet spaces opened up by the sustained, undistracted reading of a book, or by any other act of contemplation, for that matter, we make our own associations, draw our own inferences and analogies, foster our own ideas. Deep reading, as Maryanne Wolf argues, is indistinguishable from deep thinking.

If we lose those quiet spaces or fill them up with "content," we will sacrifice something important not only in ourselves but in our culture. In a recent essay, the playwright Richard Foreman eloquently described what's at stake:

I come from a tradition of Western culture, in which the ideal (my ideal) was the complex, dense and "cathedral-like" structure of the highly educated and articulate personality — a man or woman who carried inside themselves a personally constructed and unique version of the entire heritage of the West. [But now] I see within us all (myself included) the replacement of complex inner density with a new kind of self — evolving under the pressure of information overload and the technology of the "instantly available."

As we are drained of our "inner repertory of dense cultural inheritance," Foreman concluded, we risk turning into "'pancake people' — spread wide and thin as we connect with that vast network of information accessed by the mere touch of a button."

I'm haunted by that scene in *2001*. What makes it so poignant, and so weird, is the computer's emotional response to the disassembly of its mind: its despair as one circuit after another goes dark, its childlike pleading with the astronaut — "I can feel it. I can feel it. I'm afraid" — and its final reversion to what can only be called a state of innocence. HAL's outpouring of feeling contrasts with the emotionlessness that characterizes the human figures in the film, who go about their business with an almost robotic efficiency. Their thoughts and actions feel scripted, as if they're following the steps of an algorithm. In the world of *2001*, people have become so machinelike that the most human character turns out to be a machine. That's the essence of Kubrick's dark prophecy: as we come to rely on computers to mediate our understanding of the world, it is our own intelligence that flattens into artificial intelligence.

CHRIS CARROLL

High-Tech Trash

FROM *National Geographic*

JUNE IS THE WET SEASON in Ghana, but here in Accra, the capital, the morning rain has ceased. As the sun heats the humid air, pillars of black smoke begin to rise above the vast Agbogbloshie Market. I follow one plume toward its source, past lettuce and plantain vendors, past stalls of used tires, and through a clanging scrap market where hunched men bash on old alternators and engine blocks. Soon the muddy track is flanked by piles of old TVs, gutted computer cases, and smashed monitors heaped ten feet high. Beyond lies a field of fine ash speckled with glints of amber and green — the sharp broken bits of circuit boards. I can see now that the smoke issues not from one fire but from many small blazes. Dozens of indistinct figures move among the acrid haze, some stirring flames with sticks, others carrying armfuls of brightly colored computer wire. Most are children.

Choking, I pull my shirt over my nose and approach a boy of about fifteen, his thin frame wreathed in smoke. Karim says he has been tending such fires for two years. He pokes at one meditatively, and then his top half disappears as he bends into the billowing soot. He hoists a tangle of copper wire off the old tire he's using for fuel and douses the hissing mass in a puddle. With the flame-retardant insulation burned away — a process that has released a bouquet of carcinogens and other toxics — the wire may fetch a dollar from a scrap-metal buyer.

Another day in the market, on a similar ash heap above an inlet that flushes to the Atlantic after a downpour, Israel Mensah, an incongruously stylish young man of about twenty, adjusts his designer glasses and explains how he makes his living. Each day scrap sellers bring loads of old electronics — from where he doesn't know. Mensah and his partners — friends and family, including two shoeless boys raptly listening to us talk — buy a few computers or TVs. They break copper yokes off picture tubes, littering the ground with shards containing lead, a neurotoxin, and cadmium, a carcinogen that damages lungs and kidneys. They strip resalable parts such as drives and memory chips. Then they rip out wiring and burn the plastic. He sells copper stripped from one scrap load to buy another. The key to making money is speed, not safety. "The gas goes to your nose and you feel something in your head," Mensah says, knocking his fist against the back of his skull for effect. "Then you get sick in your head and your chest." Nearby, hulls of broken monitors float in the lagoon. Tomorrow the rain will wash them into the ocean.

People have always been proficient at making trash. Future archaeologists will note that at the tail end of the twentieth century, a new, noxious kind of clutter exploded across the landscape: the digital detritus that has come to be called e-waste.

More than forty years ago, Gordon Moore, cofounder of the computer-chip maker Intel, observed that computer processing power roughly doubles every two years. An unstated corollary to "Moore's law" is that at any given time, all the machines considered state of the art are simultaneously on the verge of obsolescence. At this very moment, heavily caffeinated software engineers are designing programs that will overtax and befuddle your new turbopowered PC when you try running them a few years from now. The memory and graphics requirements of Microsoft's recent Vista operating system, for instance, spell doom for aging machines that were still able to squeak by a year ago. According to the U.S. Environmental Protection Agency, an estimated 30 million to 40 million PCs will be ready for "end-of-life management" in each of the next few years.

Computers are hardly the only electronic hardware hounded by obsolescence. A switchover to digital high-definition television broadcasts is scheduled to be complete by 2009, rendering inoperable TVs that function perfectly today but receive only an analog signal. As viewers prepare for the switch, about 25 million TVs are taken out of service yearly. In the fashion-conscious mobile market, 98 million U.S. cell phones took their last call in 2005. All told, the

EPA estimates that in the United States that year, between 1.5 and 1.9 million tons of computers, TVs, VCRs, monitors, cell phones, and other equipment were discarded. If all sources of electronic waste are tallied, it could total 50 million tons a year worldwide, according to the UN Environment Programme.

So what happens to all this junk?

In the United States, it is estimated that more than 70 percent of discarded computers and monitors, and well over 80 percent of TVs, eventually end up in landfills, despite a growing number of state laws that prohibit dumping of e-waste, which may leak lead, mercury, arsenic, cadmium, beryllium, and other toxics into the ground. Meanwhile, a staggering volume of unused electronic gear sits in storage — about 180 million TVs, desktop PCs, and other components as of 2005, according to the EPA. Even if this obsolete equipment remains in attics and basements indefinitely, never reaching a landfill, this solution has its own, indirect impact on the environment. In addition to toxics, e-waste contains goodly amounts of silver, gold, and other valuable metals that are highly efficient conductors of electricity. In theory, recycling gold from old computer motherboards is far more efficient and less environmentally destructive than ripping it from the earth, often by surface-mining, which imperils pristine rain forests.

Currently, less than 20 percent of e-waste entering the solid waste stream is channeled through companies that advertise themselves as recyclers, though the number is likely to rise as states like California crack down on landfill dumping. Yet recycling, under the current system, is less benign than it sounds. Dropping your old electronic gear off with a recycling company or at a municipal collection point does not guarantee that it will be safely disposed of. While some recyclers process the material with an eye toward minimizing pollution and health risks, many more sell it to brokers who ship it to the developing world, where environmental enforcement is weak. For people in countries on the front end of this arrangement, it's a handy out-of-sight, out-of-mind solution.

Many governments, conscious that electronic waste wrongly handled damages the environment and human health, have tried to weave an international regulatory net. The 1989 Basel Convention, a 170-nation accord, requires that developed nations notify developing nations of incoming hazardous waste shipments. Environmental groups and many undeveloped nations called the terms too weak, and in 1995 protests led to an amendment known as the Basel Ban, which forbids hazardous-waste shipments to poor countries. Though the ban has yet to take effect, the European Union has written the requirements into its laws.

The EU also requires manufacturers to shoulder the burden of safe disposal. A recent EU directive encourages "green design" of electronics, setting limits for allowable levels of lead, mercury, fire retardants, and other substances. Another directive requires manufacturers to set up infrastructure to collect e-waste and ensure responsible recycling — a strategy called take-back. In spite of these safeguards, untold tons of e-waste still slip out of European ports on their way to the developing world.

In the United States, electronic waste has been less of a legislative priority. One of only three countries to sign but not ratify the Basel Convention (the other two are Haiti and Afghanistan), it does not require green design or take-back programs of manufacturers, though a few states have stepped in with their own laws. The U.S. approach, says Matthew Hale, EPA solid waste program director, is instead to encourage responsible recycling by working with industry — for instance, with a ratings system that rewards environmentally sound products with a seal of approval. "We're definitely trying to channel market forces, and look for cooperative approaches and consensus standards," Hale says.

The result of the federal hands-off policy is that the greater part of e-waste sent to domestic recyclers is shunted overseas.

"We in the developed world get the benefit from these devices," says Jim Puckett, head of Basel Action Network, or BAN, a group that opposes hazardous-waste shipments to developing nations. "But when our equipment becomes unusable, we externalize the real environmental costs and liabilities to the developing world."

Asia is the center of much of the world's high-tech manufacturing, and it is here that the devices often return when they die. China in particular has long been the world's electronics graveyard. With explosive growth in its manufacturing sector fueling demand, China's ports have become conduits for recyclable scrap of every sort: steel, aluminum, plastic, even paper. By the mid-1980s, electronic waste began freely pouring into China as well, carrying the lucrative promise of the precious metals embedded in circuit boards.

Vandell Norwood, owner of Corona Visions, a recycling company in San Antonio, Texas, remembers when foreign scrap brokers began trolling for electronics to ship to China. Today he opposes the practice, but then it struck him and many other recyclers as a win-win situation. "They said this stuff was all going to get recycled and put back into use," Norwood remembers brokers assuring him. "It seemed environmentally responsible. And it was profitable, because I was getting paid to have it taken off my hands." Huge volumes of scrap electronics were shipped out, and the profits rolled in.

Any illusion of responsibility was shattered in 2002, the year Puckett's group, BAN, released a documentary film that showed the reality of e-waste recycling in China. *Exporting Harm* focused on the town of Guiyu in Guangdong Province, adjacent to Hong Kong. Guiyu had become the dumping ground for massive quantities of electronic junk. BAN documented thousands of people — entire families, from young to old — engaged in dangerous practices like burning computer wire to expose copper, melting circuit boards in pots to extract lead and other metals, or dousing the boards in powerful acid to remove gold.

China had specifically prohibited the import of electronic waste in 2000, but that had not stopped the trade. After the worldwide publicity BAN's film generated, however, the government lengthened the list of forbidden e-wastes and began pushing local governments to enforce the ban in earnest.

On a recent trip to Taizhou, a city in Zhejiang Province south of Shanghai that was another center of e-waste processing, I saw evidence of both the crackdown and its limits. Until a few years ago, the hill country outside Taizhou was the center of a huge but informal electronics disassembly industry that rivaled Guiyu's. But these days, customs officials at the nearby Haimen and Ningbo ports — clearinghouses for massive volumes of metal scrap — are sniffing around incoming shipments for illegal hazardous waste.

High-tech scrap "imports here started in the 1990s and reached a peak in 2003," says a high school teacher whose students tested the environment around Taizhou for toxics from e-waste. He requested anonymity from fear of local recyclers angry about the drop in business. "It has been falling since 2005 and now is hard to find."

Today the salvagers operate in the shadows. Inside the open

door of a house in a hillside village, a homeowner uses pliers to rip microchips and metal parts off a computer motherboard. A buyer will burn these pieces to recover copper. The man won't reveal his name. "This business is illegal," he admits, offering a cigarette. In the same village, several men huddle inside a shed, heating circuit boards over a flame to extract metal. Outside the door lies a pile of scorched boards. In another village a few miles away, a woman stacks up bags of circuit boards in her house. She shoos my translator and me away. Continuing through the hills, I see people tearing apart car batteries, alternators, and high-voltage cable for recycling, and others hauling aluminum scrap to an aging smelter. But I find no one else working with electronics. In Taizhou, at least, the e-waste business seems to be waning.

Yet for some people it is likely too late; a cycle of disease or disability is already in motion. In a spate of studies released last year, Chinese scientists documented the environmental plight of Guiyu, the site of the original BAN film. The air near some electronics salvage operations that remain open contains the highest amounts of dioxin measured anywhere in the world. Soils are saturated with the chemical, a probable carcinogen that may disrupt endocrine and immune function. High levels of flame retardants called PBDEs — common in electronics, and potentially damaging to fetal development even at very low levels — turned up in the blood of the electronics workers. The high school teacher in Taizhou says his students found high levels of PBDEs in plants and animals. Humans were also tested, but he was not at liberty to discuss the results.

China may someday succeed in curtailing electronic waste imports. But e-waste flows like water. Shipments that a few years ago might have gone to ports in Guangdong or Zhejiang provinces can easily be diverted to friendlier environs in Thailand, Pakistan, or elsewhere. "It doesn't help in a global sense for one place like China, or India, to become restrictive," says David N. Pellow, an ethnic studies professor at the University of California, San Diego, who studies electronic waste from a social-justice perspective. "The flow simply shifts as it takes the path of least resistance to the bottom."

It is next to impossible to gauge how much e-waste is still being smuggled into China, diverted to other parts of Asia, or — increasingly — dumped in West African countries like Ghana, Nigeria, and Ivory Coast. At ground level, however, one can pick out single threads from this global toxic tapestry and follow them back to their source.

In Accra, Mike Anane, a local environmental journalist, takes me down to the seaport. Guards block us at the gate. But some truck drivers at a nearby gas station point us toward a shipment facility just up the street, where they say computers are often unloaded. There, in a storage yard, locals are opening a shipping container from Germany. Shoes, clothes, and handbags pour out onto the tarmac. Among the clutter: some battered Pentium 2 and 3 computers and monitors with cracked cases and missing knobs, all sitting in the rain. A man hears us asking questions. "You want computers?" he asks. "How many containers?"

Near the port I enter a garage-like building with a sign over the door: "Importers of British Used Goods." Inside: more age-encrusted PCs, TVs, and audio components. According to the manager, the owner of the facility imports a forty-foot container every week. Working items go up for sale. Broken ones are sold for a pittance to scrap collectors.

All around the city, the sidewalks are choked with used electronics shops. In a suburb called Darkuman, a dim stall is stacked front to back with CRT monitors. These are valueless relics in wealthy countries, particularly hard to dispose of because of their high levels of lead and other toxics. Apparently no one wants them here, either. Some are monochrome, with tiny screens. Boys will soon be smashing them up in a scrap market.

A price tag on one of the monitors bears the label of a chain of Goodwill stores headquartered in Frederick, Maryland, a forty-five-minute drive from my house. A lot of people donate their old computers to charity organizations, believing they're doing the right thing. I might well have done the same. I ask the proprietor of the shop where he got the monitors. He tells me his brother in Alexandria, Virginia, sent them. He sees no reason not to give me his brother's phone number.

When his brother Baah finally returns my calls, he turns out not to be some shady character trying to avoid the press, but a maintenance man in an apartment complex, working fifteen-hour days fixing toilets and lights. To make ends meet, he tells me, he works nights and weekends exporting used computers to Ghana through his brother. A Pentium 3 brings $150 in Accra, and he can sometimes buy the machines for less than $10 on Internet liquidation Web sites — he favors private ones, but the U.S. General Services Administration runs one as well. Or he buys bulk loads from charity stores. (Managers of the Goodwill store whose monitor ended up in Ghana denied selling large quantities of computers to dealers.) Whatever the source, the profit margin on a working computer is substantial.

The catch: nothing is guaranteed to work, and companies always try to unload junk. CRT monitors, though useless, are often part of the deal. Baah has neither time nor space to unpack and test his monthly loads. "You take it over there and half of them don't work," he says disgustedly. All you can do then is sell it to scrap people, he says. "What they do with it from that point, I don't know nothing about it."

Baah's little exporting business is just one trickle in the cataract of e-waste flowing out of the United States and the rest of the developed world. In the long run, the only way to prevent it from flooding Accra, Taizhou, or a hundred other places is to carve a new, more responsible direction for it to flow in. A Tampa, Florida, company called Creative Recycling Systems has already begun.

The key to the company's business model rumbles away at one end of a warehouse — a building-size machine operating not unlike an assembly line in reverse. "David" was what the company president, Jon Yob, called the more than $3 million investment in machines and processes when they were installed in 2006; Goliath is the towering stockpile of U.S. e-scrap. Today the machine's steel teeth are chomping up audio and video components. Vacuum pressure and filters capture dust from the process. "The air that comes out is cleaner than the ambient air in the building," vice president Joe Yob (Jon's brother) bellows over the roar. A conveyer belt transports material from the shredder through a series of sorting stations: vibrating screens of varying finenesses, magnets, a device to extract leaded glass, and an eddy current separator — akin to a reverse magnet, Yob says — that propels nonferrous metals like copper and aluminum into a bin, along with precious metals like gold, silver, and palladium. The most valuable product, shredded circuit boards, is shipped to a state-of-the-art smelter in Belgium specializing in precious-metals recycling. According to Yob, a four-foot-square box of the stuff can be worth as much as $10,000.

In Europe, where the recycling infrastructure is more developed, plant-size recycling machines like David are fairly common. So far, only three other American companies have such equipment. David can handle some 150 million pounds of electronics a year; it wouldn't take many more machines like it to process the entire country's output of high-tech trash. But under current policies, pound for pound it is still more profitable to ship waste abroad than to process it safely at home. "We can't compete economically with people who do it wrong, who ship it overseas," Joe Yob says. Creative Recycling's investment in David thus represents a gamble — one that could pay off if the EPA institutes a certification process for recyclers that defines minimum standards for the industry. Companies that rely mainly on export would have difficulty meeting such standards. The EPA is exploring certification options.

Ultimately, shipping e-waste overseas may be no bargain even for the developed world. In 2006 Jeffrey Weidenhamer, a chemist at Ashland University in Ohio, bought some cheap, Chinese-made jewelry at a local dollar store for his class to analyze. That the jewelry contained high amounts of lead was distressing, but hardly a surprise; Chinese-made leaded jewelry is all too commonly marketed in the United States. More revealing were the amounts of copper and tin alloyed with the lead. As Weidenhamer and his colleague Michael Clement argued in a scientific paper published this past July, the proportions of these metals in some samples suggest their source was leaded solder used in the manufacture of electronic circuit boards.

"The U.S. right now is shipping large quantities of leaded materials to China, and China is the world's major manufacturing center," Weidenhamer says. "It's not all that surprising things are coming full circle and now we're getting contaminated products back." In a global economy, out of sight will not stay out of mind for long.

ANDREW CURRY

Intel Inside

FROM *Wired*

FOR FIFTEEN YEARS, agents of the Stasi (short for Staatssicherheitsdienst, or State Security Service) followed her, bugged her phone and home, and harassed her unremittingly, right up until she and other dissidents helped bring down the Berlin Wall in 1989. Today the study in Ulrike Poppe's Berlin apartment is lined from floor to twelve-foot ceiling with bookshelves full of volumes on art, literature, and political science. But one shelf, just to the left of her desk, is special. It holds a pair of three-inch-thick black binders — copies of the most important documents in Poppe's secret police files. This is her Stasi shelf.

Poppe hung out with East German dissidents as a teenager, got blackballed out of college, and was busted in 1974 by the police on the thin pretext of "asocial behavior." On her way out of jail, Stasi agents asked her to be an informant, to spy on her fellow radicals, but she refused. ("I was just twenty-one, but I knew I shouldn't trust the Stasi, let alone sign anything," she says.) She went on to become a founding member of a reform-minded group called Women for Peace, and was eventually arrested thirteen more times — and imprisoned in 1983 for treason. Only an international outcry won her release.

Poppe learned to recognize many of the men assigned to tail her each day. They had crew cuts and never wore jeans or sneakers. Sometimes they took pictures of her on the sidewalk, or they piled into a white sedan and drove six feet behind her as she walked down the street. Officers waited around the clock in cars parked outside her top-floor apartment. After one of her neigh-

bors tipped her off, she found a bug drilled from the attic of the
building into the ceiling plaster of her living room.

When the Wall fell, the Stasi fell with it. The new government,
determined to bring to light the agency's totalitarian tactics, created a special commission to give victims access to their personal
files. Poppe and her husband were among the first people in Germany allowed into the archives. On January 3, 1992, she sat in front
of a cart loaded with forty binders dedicated to "Circle 2" — her
code name, it turned out. In the sixteen years since, the commission has turned up twenty more Circle 2 binders on her.

The pages amounted to a minute-by-minute account of Poppe's
life, seen from an unimaginable array of angles. Video cameras
were installed in the apartment across the street. Her friends' bedrooms were bugged and their conversations about her added to the
file. Agents investigated the political leanings of her classmates
from middle school and opened all of her mail. "They really tried
to capture everything," she says. "Most of it was just junk."

But some of it wasn't. And some of it . . . Poppe doesn't know. No
one does. Because before it was disbanded, the Stasi shredded or
ripped up about 5 percent of its files. That might not sound like
much, but the agency had generated perhaps more paper than
any other bureaucracy in history — possibly a billion pages of surveillance records, informant accounting, reports on espionage,
analyses of foreign press, personnel records, and useless minutiae.
There's a record for every time anyone drove across the border.

In the chaos of the days leading up to the actual destruction of
the Wall and the fall of East Germany's communist government,
frantic Stasi agents sent trucks full of documents to the *Papierwolfs*
and *Reisswolfs* — literally "paper-wolves" and "rip-wolves," German
for shredders. As pressure mounted, agents turned to office shredders, and when the motors burned out, they started tearing pages
by hand — 45 million of them, ripped into approximately 600 million scraps of paper.

There's no way to know what bombshells those files hide. For a
country still trying to come to terms with its role in World War II
and its life under a totalitarian regime, that half-destroyed paperwork is a tantalizing secret.

The machine-shredded stuff is confetti, largely unrecoverable.
But in May 2007, a team of German computer scientists in Berlin

announced that after four years of work, they had completed a system to digitally tape together the torn fragments. Engineers hope their software and scanners can do the job in less than five years — even taking into account the varying textures and durability of paper, the different sizes and shapes of the fragments, the assortment of printing (from handwriting to dot matrix) and the range of edges (from razor sharp to ragged and handmade). "The numbers are tremendous. If you imagine putting together a jigsaw puzzle at home, you have maybe one thousand pieces and a picture of what it should look like at the end," project manager Jan Schneider says. "We have many millions of pieces and no idea what they should look like when we're done."

As the enforcement arm of the German Democratic Republic's Communist Party, the Stasi at its height in 1989 employed 91,000 people to watch a country of 16.4 million. A sprawling bureaucracy almost three times the size of Hitler's Gestapo was spying on a population a quarter that of Nazi Germany.

Unlike the prison camps of the Gestapo or the summary executions of the Soviet Union's KGB, the Stasi strove for subtlety. "They offered incentives, made it clear people should cooperate, recruited informal helpers to infiltrate the entire society," says Konrad Jarausch, a historian at the University of North Carolina at Chapel Hill. "They beat people up less often, sure, but they psychologically trampled people. Which is worse depends on what you prefer."

That finesse helped the Stasi quell dissent, but it also fostered a pervasive and justified paranoia. And it generated an almost inconceivable amount of paper, enough to fill more than one hundred miles of shelves. The agency indexed and cross-referenced 5.6 million names in its central card catalog alone. Hundreds of thousands of "unofficial employees" snitched on friends, coworkers, and their own spouses, sometimes because they'd been extorted and sometimes in exchange for money, promotions, or permission to travel abroad.

For such an organized state, East Germany fell apart in a decidedly messy way. When the country's eastern-bloc neighbors opened their borders in the summer of 1989, tens of thousands of East Germans fled to the West through Hungary and Czechoslovakia. By autumn, protests and riots had spread throughout East Germany, with the participants demanding an end to restrictions on travel and speech. In the first week of October, thousands of demonstrators in Dresden turned violent, throwing rocks at police, who broke up the crowd with dogs, truncheons, and water cannons. The government described the thousand people they arrested as "hooligans" to state-controlled media.

But on October 9, the situation escalated. In Leipzig that night, 70,000 people marched peacefully around the city's ring road — which goes right past the Stasi office. Agents asked for permission from Berlin to break up the demonstration, but this was just a few months after the Chinese government had brutally shut down pro-democracy protests in Beijing's Tiananmen Square, to international condemnation. The East German government didn't want a similar bloodbath, so the Stasi did nothing. A week later, 120,000 people marched; a week after that, the number was 300,000 — in a city with a population of only 530,000.

In November, hundreds of East and West Berliners began dismantling the wall that bisected the city. But the communist government was still in power, negotiating with dissidents and hoping to hold on. Inside the Stasi, leaders hoped that if they weathered whatever changes were imminent, they'd be able to get back to business under a different name. But just in case, the head of the Stasi ordered the agency to start destroying the incriminating paperwork it had on hand.

In several small cities, rumors started circulating that records were being destroyed. Smoke, fires, and departing trucks confirmed the fears of angry Germans, who rushed in to their local Stasi offices, stopped the destruction, and spontaneously organized citizen committees that could post guards to secure the archives. Demonstrators spray-painted the walls with slogans like "The files belong to us" and "Stasi get out." Finally, on the evening of January 15, 1990, thousands of demonstrators pushed in the front gate of the Stasi's fortified Berlin compound.

At headquarters, agents had been more discreet than their colleagues in the hinterlands. Burning all those files would tip off angry Berliners that something was up. When the first destruction orders came in, they began stacking bags of paper in the "copper kettle," a copper-lined basement designed as a surveillance-proof computer room. The room quickly filled with bags of shredded and torn paper. Today, even the people gathering and archiving the Stasi files express grudging admiration for the achievement. "Destroying paper is shit work," says a government archivist, Stephan Wolf. "After two days your joints hurt. They ripped for two months."

But a few days after demonstrators breached the Stasi front gate, the archives still hadn't been found. A citizen group coalesced, determined to track them down. Among the searchers was a twenty-three-year-old plumber named David Gill, a democracy activist barred from university because his father was a Protestant minister. He was secretly studying theology at an underground seminary in Berlin.

Accompanied by cooperative police, Stasi agents led Gill and his compatriots through twisting alleys and concrete-walled court-yards, all eerily empty. Finally they arrived at a nondescript office building in the heart of the compound. Inside, there was more paper than he had ever imagined. "We had all lived under the pressure of the Stasi. We all knew they could know everything," Gill says today. "But we didn't understand what that meant until that moment. Suddenly it was palpable."

Gill and his crew of volunteers preserved whatever they could, commandeering trucks and borrowing cars to collect files from Stasi safe houses and storage facilities all over Berlin. Most of it was still intact. Some of it was shredded, unrecoverable. They threw that away. But then there were also bags and piles of hand-torn stuff, which they saved without knowing what to do with it. "We didn't have time to look at it all," Gill says. "We had no idea what it would mean."

Bertram Nickolay grew up in Saarland, a tiny German state close to Luxembourg that is about as far from East Germany as you could go in West Germany. He came to West Berlin's Technical University in 1974 to study engineering, the same year Ulrike Poppe was placed under Stasi surveillance on the other side of the Berlin Wall. A Christian, he felt out of place on a campus still full of leftist radicals praising East German communism and cursing the United States.

Instead, Nickolay gravitated toward exiled East German dissi-

dents and democracy activists. "I had a lot of friends who were writers and intellectuals in the GDR. There was an emotional connec-
tion," he says.

Today, Nickolay is head of the Department of Security Technology for the Fraunhofer-Institute for Production Systems and Design Technology. Fraunhofer is Europe's largest research non-
profit, with fifty-six branches in Germany alone and an annual
budget of more than $1 billion. (Fraunhofer researchers invented
the MP3 audio codec, which netted the society more than $85 mil-
lion in license fees in 2006.)

In 1996 Nickolay saw a TV news report on an unusual project. A
team working for the Stasi Records Office (BStU), the newly created ministry responsible for managing the mountain of paper left
behind by the secret police, had begun manually puzzling together
bags full of documents, scrap by scrap. The results were explosive:
here was additional proof that East Germany sheltered terrorists,
ran national sports doping programs, and conducted industrial
espionage across Western Europe. BStU's hand-reassembly program also exposed hundreds of the Stasi's secret informants —
their ranks turned out to include bishops, university professors,
and West German bureaucrats.

But the work is painfully slow. Gerd Pfeiffer, the project's manager, says he and a dwindling staff have reassembled 620,500 pages
of Stasi secrets in the thirteen years since the project began. That
works out to one bag per worker per year — 327 bags so far — and
700 years to finish.

That TV segment resonated with Nickolay — he had opposed
the East German regime, and he had the necessary technical expertise. "This is essentially a problem of automation," he says, "and
that's something Fraunhofer is very good at." He sent a letter to the
head of BStU offering his help.

The government was hesitant, but eventually the BStU issued a
proof-of-concept challenge: anyone who could digitally turn twelve
pieces of ripped-up paper into a legible document or documents
would get a grant. About twenty teams responded. Two years later,
Nickolay's group was the only one to succeed, earning a contract
for a two-year, 400-bag pilot project.

On a gray day last fall, I sat in front of two wall-mounted Sharp
Aquos flat-screen TVs hooked up to four networked computers.

Next to me, Jan Schneider, Nickolay's deputy and the manager of the Stasi document reconstruction project, booted up the machines. (This was just a demo: Nickolay refused to show me the actual lab, citing German privacy law.)

On the right-hand screen, digital images of paper fragments appeared — technicians had scanned them in using a specially designed two-camera digital imaging system. As Schneider pulled down menus and clicked through a series of descriptive choices, fragments disappeared from the screen. "Basically, we need to reduce the search space," he says. White paper or blue — or pink or green or multicolored? Plain, lined, or graph? Typewriting, handwriting, or both? Eventually, only a handful of similar-looking pieces remained. Once matched, the pieces get transferred to another processor. These pop up as a reconstructed page on the left-hand screen, rips still visible but essentially whole. (The reconstructors caught one big break: it turns out that the order-obsessed Stasi usually stuffed one bag at a time, meaning document fragments are often found together.)

Just nineteen years old when the Berlin Wall fell, Schneider doesn't share Nickolay's moral outrage. For him this is simply a great engineering challenge. He turns away from the massive monitors on the wall and picks up my business card to explain how the team is training the computers to look at these documents — the same way people do. "You see a white piece with blue writing on it — computer writing, machine writing, not handwriting — and here in the upper left is a logo. Tear it up and you'd immediately know what to look for, what goes together."

But my card is easy. For one thing, I sprang for heavy stock, and you'd be hard-pressed to tear it into enough pieces to constitute "destroyed." The Stasi files are something else entirely. In 2000 the BStU collected them and sent them to Magdeburg, a decaying East German industrial city ninety miles west of Berlin. In hand-numbered brown paper sacks, neatly stacked on row after row of steel shelves, they fill a three-story, 60,000-square-foot warehouse on the northern edge of town. Each sack contains about 40,000 fragments, for a total of 600 million pieces of paper (give or take 100 million). And each fragment has two sides. That's more than a billion images.

The numbers aren't the worst part. The documents in the bags date from the 1940s to the 1980s, and they include everything from carbon paper and newsprint to Polaroids and heavy file folders. That means the fragments have a wide variety of textures and weights. Hand-ripping stacks of thick paper creates messy, overlapping margins with a third dimension along the edges. For a computer looking for two-dimensional visual clues, overlaps show up as baffling gaps. "Keep ripping smaller and smaller and you can get pieces that are all edge," Schneider says.

The data for the 400-bag pilot project is stored on 22 terabytes' worth of hard drives, but the system is designed to scale. If work on all 16,000 bags is approved, there may be hundreds of scanners and processors running in parallel by 2010. (Right now they're analyzing actual documents, but still mostly vetting and refining the system.) Then, once assembly is complete, archivists and historians will probably spend a decade sorting and organizing. "People who took the time to rip things up that small had a reason," Nickolay says. "This isn't about revenge but about understanding our history." And not just Germany's — Nickolay has been approached by foreign officials from Poland and Chile with an interest in reconstructing the files damaged or destroyed by their own repressive regimes.

This kind of understanding isn't cheap. The German parliament has given Fraunhofer almost $9 million to scan the first 400 bags. If the system works, expanding the operation to finish the job will cost an estimated $30 million. Most of the initial cost is research and development, so the full reconstruction would mainly involve more scanners and personnel to feed the paper in.

Is it worth it? Günter Bormann, the BStU's senior legal expert, says there's an overwhelming public demand for the catharsis people find in their files. "When we started in 1992, I thought we'd need five years and then close the office," Bormann says. Instead, the Records Office was flooded with half a million requests in the first year alone. Even in cases where files hadn't been destroyed, waiting times stretched to three years. In the past fifteen years, 1.7 million people have asked to see what the Stasi knew about them.

Requests dipped in the late 1990s, but the Oscar-winning 2006 film *The Lives of Others*, about a Stasi agent who monitors a dissident playwright, seems to have prompted a surge of new applications;

2007 marked a five-year high. "Every month, 6,000 to 8,000 people decide to read their files for the first time," Bormann says. These days, the Stasi Records Office spends $175 million a year and employs 2,000 people.

This being Germany, there's even a special word for it: *Vergangenheitsbewältigung,* or "coming to terms with the past." It's not self-evident — you could imagine a country deciding, communally, to recover from a totalitarian past by simply gathering all the documents and destroying them. In fact, in 1990 the German press and citizen committees were wracked by debate over whether to do just that. Many people, however, suspected that former Stasi agents and ex-informants were behind the push to forgive and forget.

By preserving and reconstructing the Stasi archives, BStU staffers say they hope to keep history from repeating itself. In November, the first children born after the fall of the wall turned eighteen. Evidence suggests many of them have serious gaps in their knowledge of the past. In a survey of Berlin high school students, only half agreed that the GDR was a dictatorship. Two-thirds didn't know who built the Berlin Wall.

The files hold the tantalizing possibility of an explanation for the strangeness that pervaded pre-unification Germany. Even back then, Poppe wondered if the Stasi had information that would explain it all. "I always used to wish that some Stasi agent would defect and call me up to say, 'Here, I brought your file with me,'" Poppe says.

Reading the reports in that first set of forty binders spurred her to uncover as much as she could about her monitored past. Since 1995 Poppe has received eight pages from the group putting together documents by hand; the collection of taped-together paper is in a binder on her Stasi shelf.

The truth is, for Poppe the reconstructed documents haven't contained bombshells that are any bigger than the information in the rest of her file. She chooses a black binder and sets it down on the glass coffee table in her living room. After lighting a Virginia Slim, she flips to a page-long list of snitches who spied on her. She was able to match code names like Carlos, Heinz, and Rita to friends, coworkers, and colleagues in the peace movement. She even tracked down the Stasi officer who managed her case, and after she set up a sort of ambush for him at a bar — he thought he was there for a job interview — they continued to get together. Over the course of half a dozen meetings, they talked about what she had found in her files, why the Stasi was watching her, what they thought she was doing. For months, it turned out, an agent was assigned to steal her baby stroller and covertly let the air out of her bicycle tires when she went grocery shopping with her two toddlers. "If I had told anyone at the time that the Stasi was giving me flat tires, they would have laughed at me," she says. "It was a way to discredit people, make them seem crazy. I doubted my own sanity sometimes." Eventually, the officer broke off contact but continued to telephone Poppe — often drunk, often late at night, sometimes complaining about his failing marriage. He eventually committed suicide.

Poppe is looking forward to finding out what was in that last, re-constructed 5 percent. "The files were really important to see," she says, taking a drag on her cigarette and leaning forward across the coffee table. "They explained everything that happened — the letters we never got, the friends who pulled away from us. We understood where the Stasi influenced our lives, where they arranged for something to happen, and where it was simply our fault."

KEAY DAVIDSON

Blown Apart

FROM *California*

"YOUR JOB AS A SCIENTIST is to figure out how you're fooling yourself," Saul Perlmutter declares. The famed astrophysicist is sitting in the cafeteria at Lawrence Berkeley National Laboratory (LBNL), eating a falafel. Normally he talks at a machine-gun pace, but his speech, between bites, is measured. He glances out a big picture window toward the Berkeley hills and the fog-veiled universe beyond. "Our brains are . . . *so* good at seeing patterns that we sometimes see patterns that aren't there."

Perlmutter and his colleagues have spent two decades looking for patterns in the night sky — specifically, patterns in the spatial distribution of distant, dying stars that suddenly brighten and then fade. They hope to resolve an ancient puzzle: how will the universe end? Eleven years ago, in the autumn of 1997, they uncovered a big piece of the puzzle. But their discovery was so unexpected that they worried the patterns were illusory. They checked and rechecked their data, searching for some subtle error that might have misled them. A mistake would make them look like fools. But if they waited too long to report their results, rival teams might beat them to announcing the discovery and perhaps to winning a Nobel Prize.

Their shocking discovery was "dark energy," a mysterious repulsive force that apparently makes the universe expand faster and faster over time. Dark energy now threatens to undermine fundamental beliefs about physics, cosmology, perhaps even the nature of scientific discovery.

*

Ironically, Albert Einstein, who foresaw so much of modern physics and cosmology, indirectly anticipated the discovery by Perlmutter and his team. In the 1910s, Einstein proposed his general theory of relativity, which attributed the gravitational tug of masses like Earth and the sun to "warps," or curvature in what he called space-time. But he was annoyed to discover that the theory contained a hidden implication: the universe, which appeared at the time to consist of a single galaxy (the Milky Way), is either contracting or expanding. Nonsense! Anyone could look at the night sky and see that the cosmos was neither expanding nor contracting. So Einstein eliminated the implication by adding to his equations what he called "a slight modification," the infamous "cosmological constant." It was a kind of antigravitational force that, he thought, would counteract any tendency of the universe to expand or contract.

But by 1929, using evidence from a powerful new telescope, the Pasadena astronomer Edwin Hubble discovered that the universe is in fact expanding. Other galaxies are retreating from the Milky Way. Red-faced, Einstein realized that if he had trusted his original equations, he could have predicted the discovery of the expanding universe. He abandoned the cosmological constant, denouncing it as his "greatest blunder."

Based on his discovery, Hubble proposed an exciting project: an effort to forecast the long-term fate of the universe. Would it continue expanding forever? Hubble suggested that astronomers answer the question by, in effect, weighing the universe. He wanted scientists to determine the overall mass and gravity of the cosmos by using telescopes to map the distribution of galaxies (mass) across the universe. He assumed three possible fates for the universe, depending on how much mass it contained. In the first scenario, the universe is a heavyweight — massive and jam-packed with galaxies. These galaxies exert so much gravitational tug on each other that the expansion rate is quickly slowing and will eventually reverse, and the cosmos will collapse like a botched soufflé. The second possibility is that the universe is a middleweight. In that case, expansion will eventually slow to a crawl but not contract. In the final scenario, the universe is a lightweight with relatively few galaxies. Gravity isn't nearly strong enough to overcome the explosive energy of the Big Bang, so expansion will continue forever.

Hubble's goal proved much harder to achieve than expected. It was difficult to map the distances and distribution of galaxies with any reliability. Well into the 1990s, scientists were reporting widely divergent values for the present rate of cosmic expansion, the so-called Hubble constant, which is determined by analyzing how light from retreating galaxies shifts toward the red end of the spectrum. (The "redder" the shift, the faster the rate of retreat.) Still, astronomers remained convinced that if and when they found the solution, it would fit into one of Hubble's three scenarios.

To find convincing evidence for any of the expected options would have been exciting and philosophically fascinating, Perlmutter says. "What I didn't expect was that the answer would be 'none of the above.'" Thanks to him and other pioneering scientists, astronomers opted for a *fourth* option: the cosmic expansion rate is accelerating. Contrary to what almost everyone assumed in the twentieth century, the fate of the universe isn't decided solely by mass and gravity; they play second fiddle to dark energy, whatever the heck that is. Although it resembles Einstein's cosmological constant, instead of holding the universe together, the mystery force is blowing it apart.

It is hard to imagine a scientific revolutionary who looks and acts less like a rabble-rouser than Saul Perlmutter. Now nearing fifty, he's an amiable, polite, bespectacled man with thinning hair. The author and astrophysicist Donald Goldsmith described him as "Woody Allen with a Ph.D." Perlmutter was born in 1959 in Champaign-Urbana, Illinois, and grew up in Philadelphia. His father was a professor of chemical engineering at the University of Pennsylvania, and his mother was a professor of social work at Temple University. Their household was filled with talk of politics and arts. Perlmutter learned to play the violin and loved to sing. After entering Harvard as an undergraduate, he considered jointly majoring in physics and philosophy, but realized "there'd be no time for anything else."

Still, when Perlmutter arrived at Berkeley as a graduate student in physics in the early 1980s, he hoped to do research "that would address a deep philosophical question." His doctoral adviser was the physicist Richard A. Muller, who was planning to use robotic telescopes to look for supernovae and a hypothetical star called Nemesis, which Muller suspected triggered mass extinctions on

Earth by steering comets toward the inner solar system every 26 million years. Perlmutter joined that project, in which physicist Carl Pennypacker was developing a robotic telescopic search at Berkeley's Leuschner Observatory in Lafayette. Over the next few years, their hard-working robot observer detected "nearby" supernovae. Although the mystery star was never found, the supernova investigations opened the long, winding road to a historic discovery.

Perlmutter and others began trying to map the distances to supernovae of a type known as "Ia," which occur when a white dwarf star draws too much mass from its orbiting companion and explodes from the pressure. Type Ias are among the brightest supernovae, so they're visible from many billions of light years away. Unlike other supernovae, nearly all type Ias generate roughly the same amount of light. In theory, that makes them ideal for measuring distances to faraway places in the cosmos, because the apparent brightness of an object declines over distance by a predictable amount and can therefore be calculated with a simple equation. As a result, astronomers consider type Ias a "standard candle" for mapping the distribution of galaxies across the cosmos.

There were problems, though, as Perlmutter and his associates quickly learned. Some questioned whether type Ias really behave consistently. They pointed out that an astronomer who gazes upon the most distant type Ias is, in effect, gazing through a time machine — seeing the universe as it was billions of years ago. But how can we know whether the universe behaved back then as it does today? If primordial type-Ia supernovae acted differently, perhaps they are not such trustworthy markers after all.

To answer the skeptics, Perlmutter set about finding and carefully comparing as many type Ias as possible. But they occur very rarely. None had been seen in our own galaxy for several centuries, and, in even other galaxies, the short-lived supernovae are hard to spot. Undaunted, Perlmutter and Pennypacker established a collaboration with British and Australian astronomers at the Isaac Newton Telescope in the Canary Islands. By accessing the telescope's imagery via the Internet, Perlmutter was able to study new type-Ia supernovae from his office in Berkeley. In 1992 they discovered a type Ia at redshift 0.46, from a time nearly halfway back to the Big Bang — a new record. Meanwhile, new scientists gravitated to his team, by then known as the Supernova Cosmology Project (SCP).

One was Gerson Goldhaber. Then in his late sixties, Goldhaber was already one of the grand old men of Berkeley and of world physics, the offspring of a distinguished family of three generations of researchers. In the 1950s, '60s, and '70s he had participated in discoveries of new subatomic particles. By adding Goldhaber to the team, Pennypacker says, they acquired a noted physicist who was "extremely well respected by everybody and extremely reliable." With funding precarious, Pennypacker reasoned that "it would be much harder for the Lab to shut us down because Gerson was involved."

Competitors also emerged. The most important was the High-Z Supernova Search Team (z is the astronomical term for galactic redshift), founded in 1994 at Harvard. High-Z member Adam Riess later moved west to the Berkeley astronomy department, where he joined ranks with the supernova expert Alex Filippenko, formerly of the SCP team. Independently of Perlmutter's group, Riess and Filippenko began their own search for type-Ia exploding stars.

Goldhaber says he noticed something odd in 1997 while analyzing the team's twenty distant type-Ia supernovae. The objects tended to be dimmer than they should have been if cosmic theories were correct. "This worried me at the time, but not sufficiently, I'm afraid," he says. Everyone had assumed that the large-scale distribution of type Ias would show signs of a cosmic deceleration as the galaxies gravitationally tugged on each other. Incredibly, his charts revealed the deceleration was *negative*. Eleven years later, Goldhaber recalls the discovery with enthusiasm: "Negative deceleration means acceleration!" The retreat of the galaxies is speeding up. Some mysterious force, a kind of antigravity, is counteracting the gravitational pull of cosmic matter.

After "many checks and rechecks" of his data, Goldhaber announced his analysis of thirty-eight type-Ia supernovae at the SCP team's weekly meeting on September 24, 1997. He advised his colleagues to double-check his results: "I've been known to make mistakes," he acknowledged — a comment that was italicized in the meeting notes. By November, he says, "everyone became convinced that this astonishing result was indeed correct." In December Perlmutter presented the team's preliminary findings in speeches at Berkeley and Santa Cruz. The following year, 1998, *Science* magazine named the discovery of the expanding universe "Breakthrough of the Year."

But does it make sense to say that Goldhaber — or, for that matter, any single person "discovered" dark energy at a specific time? Members of the various teams disagree on who discovered what and when. This is typical in science, as the often-numerous participants in a historic discovery review their actions and debate whose contributions were decisive. There's plenty at stake — awards, for example. So far the many awards given to dark-energy researchers have usually been distributed among figures from the High-Z and SCP teams. There is no doubt that Perlmutter's team collected much of the data on what later turned out to be dark energy, but some members of other teams feel they also deserve credit for independently analyzing the data and because there was some collaboration among the groups. Even within Perlmutter's group, there is disagreement over Goldhaber's claim that he presented the first clear evidence for a cosmological constant, a.k.a. dark energy, in September 1997. To date the controversy has been muted, in part because many of these scientists still work together.

The fact is, scientific breakthroughs often occur through the toil of dozens, hundreds, even thousands of workers, from Ph.D.s to lab techs, sometimes over many decades. Astronomy has long been seen as a "small science," but that's fast changing: as a discipline, it is increasingly dependent on big, expensive research projects bankrolled by the federal government. This autumn, for example, in a joint collaboration, NASA and the Department of Energy are expected to formally request proposals for a future satellite that will explore dark energy. Perlmutter and his colleagues propose building a satellite called SNAP, the Supernova Acceleration Probe. The project will be "Big Science" incarnate, costing hundreds of millions of dollars.

With his colleague Michael Levi, Perlmutter took me to see a full-scale wooden model of the probe inside a large machine shop on a hillside at LBNL. We donned construction helmets to enter the building. The model is a precise replica of SNAP, right down to its complex astronomical mirror and its half-billion-pixel electronic imager. Patting one of its struts, Perlmutter acknowledged that a few skeptics have cautioned that dark energy might be so subtle an effect that all efforts to explain it will fail. Despite its cost, there is no guarantee the project will reveal the underlying nature of dark energy.

There is no known physical explanation for dark energy — indeed, no one has a clue what dark energy is; it defies all existing paradigms. But Perlmutter defends SNAP and other projects that seek to understand it. "Until you look, you don't know." Still, he says skepticism is vital to scientific progress: "You spend 95 percent of your time looking for every possible way that you could be wrong . . . That's why you get the 'Mr. Spock' characterization of the unenthusiastic, wet-blanket scientist, because there *is* the absolute need to be your own worst skeptic."

Because scientific progress now requires multiple and varied contributors, Perlmutter believes the question of who discovered dark energy and when is ultimately an empty one. Science is a group enterprise; and because groups change their collective minds more slowly than individuals, historic discoveries usually unfold over months or years, not moments. "There's no 'aha!' moment," he says. "It's more like ahhhhhhhhhh . . . haaaaaaaaaa!"

DOUGLAS FOX

Did Life Begin in Ice?

FROM *Discover*

ONE MORNING IN LATE 1997, Stanley Miller lifted a glass vial from a cold, bubbling vat. For twenty-five years he had tended the vial as though it were an exotic orchid, checking it daily, adding a few pellets of dry ice as needed to keep it at −108 degrees Fahrenheit. He had told hardly a soul about it. Now he set the frozen time capsule out to thaw, ending the experiment that had lasted for more than one-third of his sixty-eight years.

Miller had filled the vial in 1972 with a mixture of ammonia and cyanide, chemicals that scientists believe existed on early Earth and may have contributed to the rise of life. He had then cooled the mix to the temperature of Jupiter's icy moon Europa — too cold, most scientists had assumed, for much of anything to happen. Miller disagreed. Examining the vial in his laboratory at the University of California at San Diego, he was about to see who was right.

As Miller and his former student Jeffrey Bada brushed the frost from the vial that morning, they could see that something had happened. The mixture of ammonia and cyanide, normally colorless, had deepened to amber, highlighting a web of cracks in the ice. Miller nodded calmly, but Bada exclaimed in shock. It was a color that both men knew well — the color of complex polymers made up of organic molecules. Tests later confirmed Miller and Bada's hunch. Over a quarter-century, the frozen ammonia-cyanide blend had coalesced into the molecules of life: nucleobases, the building blocks of RNA and DNA, and amino acids, the building blocks of proteins. The vial's contents would support a new account of how life began on Earth and would arouse both surprise and skepticism around the world.

Although one of Miller's final experiments, it certainly wasn't the final word. The last several years have seen a steady stream of corroborating evidence, including one experiment — so new it has not yet been published — that Miller's colleague, the late Leslie Orgel, called "astonishing."

For decades, those studying the origin of life have imagined that it emerged in balmy conditions from primordial soups, tropical ponds, even boiling volcanic vents. Miller and a few other scientists began to suspect that life began not in warmth but in ice — at temperatures that few living things can now survive. The very laws of chemistry may have favored ice, says Bada, now at the Scripps Institution of Oceanography in La Jolla, California. "We've been arguing for a long time," he says, "that cold conditions make much more sense, chemically, than warm conditions."

Miller's frozen experiment is a striking testament to the idea. Although life requires liquid water, small amounts of liquid can persist even at −60°F. Microscopic pockets of water within the ice may have gathered simple molecules like the ones Miller synthesized, assembling them into longer and longer chains. A single cubic yard of sea ice contains a million or more liquid compartments, microscopic test tubes that could have created unique mixtures of RNA that eventually formed the first life.

If life on Earth arose from ice, then our chances of finding life elsewhere in the solar system — not to mention elsewhere in the galaxy — may be better than we ever imagined.

The vial of ammonia and cyanide chilling in Miller's lab was just one of the chemical cocktails he kept, aging like wine in a cellar. Some of the samples sat in freezers, others under the sink, and still others in water baths maintained at various temperatures. They were part of an effort to understand chemical reactions that must have unfolded over millennia on early Earth. The location of every sample was stored in Miller's head; occasionally he would give one to a student to analyze.

Matthew Levy, once a graduate student of Miller's and now a molecular biologist at the Albert Einstein College of Medicine in New York City, recalls being handed one of the twenty-five-year-old samples to work on. "I was scared," he says. "I was thinking, these samples are older than I am." Levy burned holes in his shirts over the next few weeks as he dissolved the samples with hydrochloric acid and ran them through an instrument called a high-performance liquid chromatograph to identify the chemicals that had formed. Red and green pens on the device traced out telltale peaks on a scrolling strip of paper. Those peaks corresponded to seven different amino acids and eleven types of nucleobases.

"What was remarkable," Bada says, "is that the yield in these frozen experiments was better, for some compounds, than it was with room-temperature experiments."

There were people who found the results a little *too* remarkable. When Bada and Miller submitted their findings to a top-tier science journal, the article was rejected. A reviewer of the manuscript felt that those molecules must surely have formed while the samples were thawing, not while frozen at the ridiculously low temperature of −108°F. So Miller, Bada, and Levy did more experiments to show that thawing played no role. They published their results in another journal, *Icarus,* in 2000.

The skepticism they faced was understandable. Chemical reactions do slow down as the temperature drops, and according to standard calculations, the reactions that assemble cyanide molecules into amino acids and nucleobases should run 100,000 times more slowly at −112°F than at room temperature. By that reckoning, even if Miller had run his experiment for two hundred and fifty years — let alone twenty-five — he should have seen nothing.

That is the main argument against Miller's experiment and against a cold origin of life in general. But strange things happen when you freeze chemicals in ice. Some reactions slow down, but others actually speed up — especially reactions that involve joining small molecules into larger ones. This seeming paradox is caused by a process called eutectic freezing. As an ice crystal forms, it stays pure: only molecules of water join the growing crystal, while impurities like salt or cyanide are excluded. These impurities become crowded in microscopic pockets of liquid within the ice, and this crowding causes the molecules to collide more often. Chemically speaking, it transforms a tepid seventh-grade school dance into a raging molecular mosh pit.

"Usually as you cool things, the reaction rates go down," concluded Leslie Orgel, who studied the origins of life at the Salk Institute in La Jolla from the 1960s until his death last October. "But with eutectic freezing, the concentrations go up so fast that they more than make up" for the difference.

Cyanide is a good candidate as a precursor molecule in the life-in-a-freezer model for several reasons. First, planetary scientists suspect that cyanide was abundant on early Earth, deposited here by comets or created in the atmosphere by ultraviolet light or by lightning (once the atmosphere became oxygen rich, 2.5 billion years ago, the process would have stopped). Second, although cyanide is lethal to modern animals, it has a convenient tendency to self-assemble into larger molecules. Third, and perhaps most important, no matter how much cyanide rained down, it could become concentrated only in a cold environment — not in warm coastal lagoons — because it evaporates more quickly than water.

"The strong point of freezing," according to Orgel, "is that you concentrate things very efficiently without evaporation." Freezing also helps preserve fragile molecules like nucleobases, extending their lifetime from days to centuries and giving them time to accumulate and perhaps organize into something more interesting — like life.

Orgel and his coworkers proposed these ideas in 1966, when he showed that frozen cyanide efficiently assembles into larger molecules. Alan Schwartz, a biochemist at the University of Nijmegen in the Netherlands, took the idea further when he showed in 1982 that frozen cyanide, in the presence of ammonia, can form a nucleobase called adenine. And Stanley Miller likely had the eutectic effect in mind when he stowed his now famous samples in a freezing chamber full of dry ice and acetone.

While Miller and Orgel followed their clues in the lab, other scientists pursued their obsession with life's chilly origins to the ends of the earth.

In July 2002 a small skiff dropped Hauke Trinks on the beach of Nordaustland, a rocky island encased in glaciers and nearly devoid of plants. Trinks, then a physicist at the Technical University of Hamburg-Harburg in Germany, had come to Nordaustland — far north of the Arctic Circle — to peer 4 billion years back in time to an era shortly after the end of the bombardment of Earth by aster-

oids. According to some solar evolution models, the sun was some
30 percent dimmer at that time, providing less heat to Earth. So as
soon as the hail of asteroids stopped, Earth may have cooled to an
average surface temperature of −40°F, and a crust of ice as much as
1,000 feet thick may have covered the oceans. Many scientists have
puzzled over how life could have arisen on a planet that was essentially a giant snowball. The answer, Trinks suspected, involved sea
ice.

Trinks had become interested in sea ice ten years before, while
studying its tendency to accumulate pollutants from the atmosphere and concentrate them in liquid pockets within the ice. He
set out to explore whether a layer of ice covering early Earth's
oceans might have gathered and assembled organic molecules.

With a few crates of supplies and two sled dogs, Trinks and his
partner, Marie Tieche, hunkered down in a cabin on Nordaustland
for thirteen months. Each morning they monitored the temperature of the ice and prepared the day's experiments. To study the
networks of liquid pockets, Trinks injected dyes into the ice and
watched through a microscope as they spread.

Winter deepened, twenty-four-hour darkness descended, and
the mercury plummeted to −20°F. Trinks continued his experiments, sometimes banging pans together to chase polar bears away.
Once a walrus lunged up through the ice and dragged several of
Trinks's instruments into the ocean.

He built a makeshift lab table from planks of wood and discarded gasoline cans. He examined slices of sea ice under the microscope, his hood pulled tight around his eyes. Turning a knob
with a gloved hand, he nudged a metal electrode nearly as fine as a
red blood cell closer to an ice crystal. The needle on his voltmeter
jerked sideways, registering a sharp drop in voltage on the crystal's
surface — evidence of a microscopic electric field that might arrange and orient molecules on the ice's surface.

By the time Trinks returned to Hamburg in 2003, he had formu-
lated a theory that ice was doing much more than just concentrat-
ing chemicals. The ice surface is a checkerboard of positive and
negative charges; he imagined those charges grabbing individual
nucleobases and stacking them like Pringles in a can, helping them
coalesce into a chain of RNA. "The surface layer between ice and
liquid is very complicated," he says. "There is strong bonding be-

tween the surface of the ice and the liquid. Those bondings are important for producing long organic chains like RNA."

At a lecture in Hamburg in 2003, Trinks met up with the chemist Christof Biebricher, who was studying how the first RNA chains could have formed in the absence of the enzymes that guide their formation in living cells. Trinks approached Biebricher with his sea-ice theory, but to Biebricher, the experiments to test it sounded messy — more like a margarita recipe than a serious scientific investigation. "Chemists," says Biebricher, "do not like heterogeneous substances like ice." But Trinks convinced him to try it in his laboratory at the Max Planck Institute for Biophysical Chemistry in Göttingen, Germany.

Biebricher sealed small amounts of RNA nucleobases — adenine, cytosine, guanine — with artificial seawater into thumb size plastic tubes and froze them. After a year, he thawed the tubes and analyzed them for chains of RNA.

For decades researchers had tried to coax RNA chains to form under all sorts of conditions without using enzymes; the longest chain formed, which Orgel accomplished in 1982, consisted of about forty nucleobases. So when Biebricher analyzed his own samples, he was amazed to see RNA molecules up to four hundred bases long. In newer, unpublished experiments, he says he has observed RNA molecules seven hundred bases long. Biebricher's results are so fantastic that some colleagues have wondered whether accidental contamination played a role. Orgel defended the work. "It's a remarkable result," he said. "It's so remarkable that everyone wants better evidence than they would for an unremarkable result. But I think it's right."

Biebricher had loaded the deck somewhat, because he wasn't growing RNA chains from nothing. Before he froze his samples, he added an RNA template — a single-strand chain of RNA that guides the formation of a new strand of RNA. As that new RNA strand grows, it adheres to the template like one half of a zipper to the other. This must be how the first genes, made of RNA, would have copied themselves. But the first step was the formation of the original RNA molecule that served as a template, and how that step happened remains a mystery.

Ice may prove the crucial ingredient here, too. David W. Deamer and his former student Pierre-Alain Monnard (now at Los Alamos

National Laboratory in New Mexico) have run experiments frozen at 0°F for a month without the aid of templates. In those relatively brief experiments they already see RNA molecules up to thirty bases long, at least as long as other researchers have seen in similar experiments without ice.

That is a good start, but it leaves unanswered the question of how you get from tiny snippets of RNA to longer, well-crafted chains that could have acted as the first enzymes, doing fancy things like copying themselves. The shortest RNA enzyme chains known today are about fifty bases long; most have more than one hundred. To work effectively, moreover, an RNA enzyme must fold correctly, which requires exactly the right sequence of bases.

A young scientist named Svetlana Balatskaya stumbled upon a possible answer. She was working at SomaGenics, a biotech company in Santa Cruz, California, to develop RNA enzymes that silence a gene called tumor necrosis factor-alpha, which is involved in rheumatoid arthritis. Her RNA enzymes were behaving strangely: they normally consisted of a single segment of RNA, but every time she cooled them below freezing to purify them, the chain of RNA spontaneously joined its ends into a circle, like a snake biting its tail. Brian Johnston and Sergie Kazakov, the head scientists at SomaGenics, soon noticed that another RNA enzyme, called hairpin ribozyme, also acted strangely. At room temperature, hairpin acts like reversible scissors, alternately snipping other RNA molecules into pieces and then reattaching them, in equal proportion. But when they froze hairpin, it suddenly ran in only one direction: it glued other RNA chains together end to end.

Kazakov and Johnston realized that the ice was driving both enzymes to work in one direction. Normally when an enzyme cuts an RNA chain in two, water molecules form tight shells around the newly formed ends. These unwieldy cages prevent the ends of the RNA chains from coming close enough to reattach. By removing most of the liquid water, preventing those cages from forming and also causing the RNA enzyme to pucker into a slightly different shape, the ice creates conditions that allow the RNA enzyme to work in just one direction, joining RNA chains.

The SomaGenics scientists wondered whether an icy spot on early Earth could have driven a primitive enzyme to do the same. To investigate this, Alexander Vlassov, another scientist at Soma-

Genics, introduced random mutations into the hairpin ribozyme, shortened it from its normal length of fifty-eight bases, and even cut it into pieces — all in an effort to produce RNA enzymes that were as dodgy and imperfect as early Earth's first enzymes likely were. These pseudoprimitive RNA enzymes do nothing at room temperature. But freeze them and they become active, joining other RNA molecules at a slow but measurable rate.

These findings inspired a theory that the first, extremely inefficient RNA enzymes got help from ice, which created an environment that encouraged short segments of RNA to stick together and behave like a single larger RNA molecule. "Freezing stabilizes the complexes formed from multiple pieces of RNA," concludes Kazakov. "So small pieces of RNA could be enzymes, not just large fifty-base molecules."

Equally telling, the pseudoprimitive RNA enzymes that Vlassov made grabbed and joined just about any other molecule. Enzymes on early Earth might have done the same, joining random segments of five or ten RNA bases to form a variety of sequences.

All these processes would occur in microscopic pockets of liquid within the ice. "You have billions and billions of different possibilities," Trinks says, "because you have billions of these small channels," each like a microscopic test tube containing a unique RNA experiment. On the young Earth, pockets of liquid could have expanded into a network of channels that mixed their contents during freeze-thaw cycles, like day-night temperature changes in summer. In winter the liquid pores would have contracted and become isolated again, returning to their separate experiments. With all the mixing, something special might eventually have formed: an RNA molecule that made rough copies of itself. And as Earth warmed, these molecules might have found a home in newly thawed seas or ponds, where something even more complex might have emerged — such as a cell-like membrane. "You have something that is multiplying itself, and you have variation that is inherited," says Antonio Lazcano, a biology researcher and professor at the National Autonomous University of Mexico, in Mexico City. "There you have the onset of Darwinian evolution. I'm willing to call that living."

No one can really know if this is how life began. Other theories posit that mineral surfaces organized key molecules or volcanic sources synthesized amino acids. These theories need not be mutually exclusive. Glaciers on early Earth could have scooped up mineral dust; volcanoes could have rained ash onto nearby sea ice. Primordial ice "must have been full of impurities," Lazcano says, "and those impurities must have had catalytic effects, enhancing the synthesis or destruction of some compounds."

Shortly after Miller finished his twenty-five-year experiment, he suffered a stroke that ended his career. His laboratory, with forty years of samples, was emptied in 2002 to make way for a building renovation. Experiments that had run for years or decades were discarded without ever being analyzed. As Bada rescued a few items from his mentor's freezer, safety personnel stood by in hazmat suits, sent by university officials concerned about rumors of toxic cyanide. Any sample that couldn't be identified was incinerated. Miller was present for a few hours of this ordeal, struggling to find words to identify the vials that he had known so well.

Miller died on May 20, 2007, but the provocative theory he helped nurture lives on. In the latest twist, Miller's ideas are influencing not just theories about life's origin on Earth but also investigations about the potential for life elsewhere in the solar system. In fact, it was a dinner conversation with Bada regarding Jupiter's moon Europa that prompted Miller to open his twenty-five-year-old samples back in 1997. While most scientists were focusing on the possibility of life in Europa's ocean, he and Bada had been talking about what biochemistry might happen in the ten-mile-thick layer of ice atop the ocean. Those speculations are more relevant than ever, with recent discoveries of geysers on Saturn's icy moon Enceladus and elaborate organic molecules on Titan, another Saturnian moon. Recent studies show that Mars too has vast quantities of buried ice, especially at its poles.

If life arose in one of these frozen zones, it might still exist there. Although life as we know it requires liquid water, there are places where life survives well below freezing. In the microscopic veins that permeate Arctic ice, for example, the high concentration of salt can maintain traces of water in a liquid state down to $-65°F$. Bacteria and diatoms inhabit those liquid veins, and Hajo Eicken, a glaciologist at the University of Alaska at Fairbanks, suspects that similar habitats could exist in the lower, warmer layers of ice on

Europa, and perhaps on the other moons as well. "There's potentially hundreds of meters of ice, if not maybe a few kilometers, that may well be quite habitable," Eicken says.

Liquid water — and life — occurs in other cold places, too. Films of liquid water persist far below freezing, like coatings of condensation, on the surfaces of some minerals. Under some conditions, these films may stay liquid down to $-90°F$. Bacteria beneath films of liquid water only several molecules thick have been found clinging to microscopic grains of clay in ice cores from Greenland. Slowly consuming the iron in a single grain, these bacteria could get by for a million years before exhausting their food supply; at colder temperatures, where metabolic demands are lower, they might survive for hundreds of millions of years.

If life arose in ice on Earth, then why not on Mars, Europa, or Enceladus? "You've got to keep an open mind in this business," Bada says "If I were going to make a bet about what we'd find if we discover life elsewhere in the universe, I would suspect it would be more cold-adapted than hot-adapted."

ADAM FRANK

The Day Before Genesis

FROM *Discover*

FOR PAUL STEINHARDT AND NEIL TUROK, the Big Bang ended on a summer day in 1999 in Cambridge, England. Sitting together at a conference they had organized, called "A School on Connecting Fundamental Physics and Cosmology," the two physicists suddenly hit on the same idea. Maybe science was finally ready to tackle the mystery of what made the Big Bang go bang. And if so, then maybe science could also address one of the deepest questions of all: what came before the Big Bang?

Steinhardt and Turok — working closely with a few like-minded colleagues — have now developed these insights into a thorough alternative to the prevailing, Genesis-like view of cosmology. According to the Big Bang theory, the whole universe emerged during a single moment some 13.7 billion years ago. In the competing theory, our universe generates and regenerates itself in an endless cycle of creation. The latest version of the cyclic model even matches key pieces of observational evidence supporting the older view.

This is the most detailed challenge yet to the 40-year-old orthodoxy of the Big Bang. Some researchers go further and envision a type of infinite time that plays out not just in this universe but in a multiverse — a multitude of universes, each with its own laws of physics and its own life story. Still others seek to revise the very idea of time, rendering the concept of a "beginning" meaningless.

All of these cosmology heretics agree on one thing: the Big Bang no longer defines the limit of how far the human mind can explore.

Big Idea 1: The Incredible Bulk

The latest elaboration of Steinhardt and Turok's cyclic cosmology, spearheaded by Evgeny Buchbinder of Perimeter Institute for Theoretical Physics in Waterloo, Ontario, was published last December. Yet the impulse behind this work far predates modern theories of the universe. In the fourth century A.D., St. Augustine pondered what the Lord was doing before the first day of Genesis (wryly repeating the exasperated retort that "He was preparing Hell for those who pry too deep"). The question became a scientific one in 1929, when Edwin Hubble determined that the universe was expanding. Extrapolated backward, Hubble's observation suggested the cosmos was flying apart from an explosive origin, the fabled Big Bang.

In the standard interpretation of the Big Bang, which took shape in the 1960s, the formative event was not an explosion that occurred at some point in space and time — it was an explosion *of* space and time. In this view, time did not exist beforehand. Even for many researchers in the field, this was a bitter pill to swallow. It is hard to imagine time just starting: how does a universe decide when it is time to pop into existence?

For years, every attempt to understand what happened in that formative moment quickly hit a dead end. In the standard Big Bang model, the universe began in a state of near-infinite density and temperature. At such extremes the known laws of physics break down. To push all the way back to the beginning of time, physicists needed a new theory, one that blended general relativity with quantum mechanics.

The prospects for making sense of the Big Bang began to improve in the 1990s as physicists refined their ideas in string theory, a promising approach for reconciling the relativity and quantum views. Nobody knows yet whether string theory matches up with the real world — the Large Hadron Collider, a particle smasher coming on line later this year, may provide some clues — but it has already inspired stunning ideas about how the universe is constructed. Most notably, current versions of string theory posit seven hidden dimensions of space in addition to the three we experience.

Strange and wonderful things can happen in those extra dimensions: That is what inspired Steinhardt (of Princeton University) and Turok (of Cambridge University) to set up their fateful conference in 1999. "We organized the conference because we both felt that the standard Big Bang model was failing to explain things," Turok says. "We wanted to bring people together to talk about what string theory could do for cosmology."

The key concept turned out to be a "brane," a three-dimensional world embedded in a higher-dimensional space (the term, in the language of string theory, is just short for *membrane*). "People had just started talking about branes when we set up the conference," Steinhardt recalls. "Together Neil and I went to a talk where the speaker was describing them as static objects. Afterward we both asked the same question: What happens if the branes can move? What happens if they collide?"

A remarkable picture began to take shape in the two physicists' minds. A sheet of paper blowing in the wind is a kind of two-dimensional membrane tumbling through our three-dimensional world. For Steinhardt and Turok, our entire universe is just one sheet, or 3-D brane, moving through a four-dimensional background called "the bulk." Our brane is not the only one; there are others moving through the bulk as well. Just as two sheets of paper could be blown together in a storm, different 3-D branes could collide within the bulk.

The equations of string theory indicated that each 3-D brane would exert powerful forces on others nearby in the bulk. Vast quantities of energy lie bound up in those forces. A collision between two branes could unleash those energies. From the inside, the result would look like a tremendous explosion. Even more intriguing, the theoretical characteristics of that explosion closely matched the observed properties of the Big Bang — including the cosmic microwave background, the afterglow of the universe's fiercely hot early days. "That was amazing for us because it meant colliding branes could explain one of the key pieces of evidence people use to support the Big Bang," Steinhardt says.

Three years later came a second epiphany: Steinhardt and Turok found their story did not end after the collision. "We weren't looking for cycles," Steinhardt says, "but the model naturally produces them." After a collision, energy gives rise to matter in the brane worlds. The matter then evolves into the kind of universe we know: galaxies, stars, planets, the works. Space within the branes expands, and at first the distance between the branes (in the bulk) grows too. When the brane worlds expand so much that their space is nearly empty, however, attractive forces between the branes draw the world-sheets together again. A new collision occurs, and a new cycle of creation begins. In this model, each round of existence — each cycle from one collision to the next — stretches about a trillion years. By that reckoning, our universe is still in its infancy, being only 0.1 percent of the way through the current cycle.

The cyclic universe directly solves the problem of before. With an infinity of Big Bangs, time stretches into forever in both directions. "The Big Bang was not the beginning of space and time," Steinhardt says. "There was a before, and before matters because it leaves an imprint on what happens in the next cycle."

Not everyone is pleased by this departure from the usual cosmological thinking. Some researchers consider Steinhardt and Turok's ideas misguided or even dangerous. "I had one well-respected scientist tell me we should stop because we were undermining public confidence in the Big Bang," Turok says. But part of the appeal of the cyclic universe is that it is not just a beautiful idea — it is a testable one.

The standard model of the early universe predicts that space is full of gravitational waves, ripples in space-time left over from the first instants after the Big Bang. These waves look very different in the cyclic model, and those differences could be measured — as soon as physicists develop an effective gravity-wave detector. "It may take 20 years before we have the technology," Turok says, "but in principle it can be done. Given the importance of the question, I'd say it's worth the wait."

Big Idea 2: Time's Arrow

While the concept of a cyclic universe provides a way to explore the Big Bang's past, some scientists believe that Steinhardt and Turok have skirted the deeper issue of origins. "The real problem is not the beginning of time but the arrow of time," says Sean Carroll, a theoretical physicist at Caltech. "Looking for a universe that repeats itself is exactly what you do not want. Cycles still give us a time that flows with a definite direction, and the direction of time is the very thing we need to explain."

In 2004 Carroll and a graduate student of his, Jennifer Chen, came up with a much different answer to the problem of before. In his view, time's arrow and time's beginning cannot be treated separately: there is no way to address what came before the Big Bang until we understand why the before precedes the after. Like Steinhardt and Turok, Carroll thinks that finding the answer requires rethinking the full extent of the universe, but Carroll is not satisfied with adding more dimensions. He also wants to add more universes — a whole lot more of them — to show that, in the big picture, time does not flow so much as advance symmetrically backward and forward.

The one-way progression of time, always into the future, is one of the greatest enigmas in physics. The equations governing individual objects do not care about time's direction. Imagine a movie of two billiard balls colliding; there is no way to say if the movie is being run forward or backward. But if you gather a zillion atoms together in something like a balloon, past and future look very different. Pop the balloon and the air molecules inside quickly fill the entire space; they never race backward to reinflate the balloon.

In any such large group of objects, the system trends toward equilibrium. Physicists use the term *entropy* to describe how far a system is from equilibrium. The closer it is, the higher its entropy; full equilibrium is, by definition, the maximum value. So the path from low entropy (all the molecules in one corner of the room, unstable) to maximum entropy (the molecules evenly distributed in the room, stable) defines the arrow of time. The route to equilibrium separates before from after. Once you hit equilibrium the arrow of time no longer matters, because change is no longer possible.

"Our universe has been evolving for 13 billion years," Carroll says, "so it clearly did not start in equilibrium." Rather, all the matter, energy, space, and even time in the universe must have started in a state of extraordinarily low entropy. That is the only way we could begin with a Big Bang and end up with the wonderfully diverse cosmos of today. Understand how that happened, Carroll argues, and you will understand the bigger process that brought our universe into being.

To demonstrate just how strange our universe is, Carroll considers all the other ways it might have been constructed. Thinking about the range of possibilities, he wonders: "Why did the initial setup of the universe allow cosmic time to have a direction? There are an infinite number of ways the initial universe could have been set up. An overwhelming majority of them have high entropy." These high-entropy universes would be boring and inert; evolution and change would not be possible. Such a universe could not produce galaxies and stars, and it certainly could not support life.

It is almost as if our universe were fine-tuned to start out far from equilibrium so it could possess an arrow of time. But to a physicist, invoking fine-tuning is akin to saying "a miracle occurred." For Carroll, the challenge was finding a process that would explain the universe's low entropy naturally, without any appeal to incredible coincidence or (worse) to a miracle.

Carroll found that process hidden inside one of the strangest and most exciting recent elaborations of the Big Bang theory. In 1984, MIT physicist Alan Guth suggested that the very young universe had gone through a brief period of runaway expansion, which he called "inflation," and that this expansion had blown up one small corner of an earlier universe into everything we see. In the late 1980s Guth and other physicists, most notably Andrei Linde, now at Stanford, saw that inflation might happen over and over in a process of "eternal inflation." As a result, pocket universes much like our own might be popping out of the uninflated background all the time. This multitude of universes was called, inevitably, the multiverse.

Carroll found in the multiverse concept a solution to both the direction and the origin of cosmic time. He had been musing over the arrow of time as far back as graduate school in the late 1980s, when he published papers on the feasibility of time travel using known physics. Eternal inflation suggested that it was not enough to think about time in our universe only; he realized he needed to consider it in a much bigger, multiverse context.

"We wondered if eternal inflation could work in both directions," Carroll says. "That means there would be no need for a single Big Bang. Pocket universes would always sprout from the uninflated background. The trick needed to make eternal inflation work was to find a generic starting point: an easy-to-achieve condition that would occur infinitely many times and allow eternal inflation to flow in both directions."

A full theory of eternal inflation came together in Carroll's mind in 2004, while he was attending a five-month workshop on cosmology at the University of California at Santa Barbara's famous Kavli

Institute of Theoretical Physics with his student Jennifer Chen. "You go to a place like Kavli and you are away from the normal responsibilities of teaching," Carroll says. "That gives you time to pull things together." In those few months, Carroll and Chen worked out a vision of a profligate multiverse without beginnings, endings, or an arrow of time.

"All you need," Carroll says, with a physicist's penchant for understatement, "is to start with some empty space, a shard of dark energy, and some patience." Dark energy — a hidden type of energy embedded in empty space, whose existence is strongly confirmed by recent observations — is crucial because quantum physics says that any energy field will always yield random fluctuations. In Carroll and Chen's theory, fluctuations in the dark-energy background function as seeds that trigger new rounds of inflation, creating a crop of pocket universes from empty space.

"Some of these pocket universes will collapse into black holes and evaporate, taking themselves out of the picture," Carroll says. "But others will expand forever. The ones that expand eventually thin out. They become the new empty space from which more inflation can start." The whole process can happen again and again. Amazingly, the direction of time does not matter in the process. "That is the funny part. You can evolve the little inflating universes in either direction away from your generic starting point," Carroll says. In the super-far past of our universe, long before the Big Bang, there could have been other Big Bangs for which the arrow of time ran in the opposite direction.

On the grandest scale, the multiverse is like a foam of interconnected pocket universes, completely symmetric with respect to time. Some universes move forward, but overall, an equal number move backward. With infinite space in infinite universes, there are no bounds on entropy. It can always increase; every universe is born with room (and entropy) to evolve. The Big Bang is just *our* Big Bang, and it is not unique. The question of before melts away because the multiverse has always existed and always will, evolving but — in a statistical sense — always the same.

After completing his multiverse paper with Chen, Carroll felt a twinge of dismay. "When you finish something like this, it's bittersweet. The fun with hard problems can be in the chase," he says. Luckily for him, the chase goes on. "Our paper really expresses a minority viewpoint," he admits. He is now hard at work on follow-up papers fleshing out the details and bolstering his argument.

Big Idea 3: The Nows Have It

In 1999, while Steinhardt and Turok were convening in Cambridge and Carroll was meditating on the meaning of the multiverse, rebel physicist Julian Barbour published *The End of Time* — a manifesto suggesting that attempts to address what came before the Big Bang were based on a fundamental mistake. There is no need to find a solution to time's beginning, Barbour insisted, because time does not actually exist.

Back in 1963, a magazine article had changed Barbour's life. At the time he was just a young physics graduate student heading off for a relaxing trip to the mountains. "I was studying in Germany and had brought an article with me on holiday to the Bavarian Alps," says Barbour, now 71. "It was about the great physicist Paul Dirac. He was speculating on the nature of time and space in the theory of relativity." After finishing the article Barbour was left with a question he would never be able to relinquish: what, really, is time? He could not stop thinking about it. He turned around half-way up the mountain and never made it to the top.

"I knew that it would take years to understand my question," Barbour recalls. "There was no way I could have a normal academic career, publishing paper after paper, and really get anywhere." With bulldog determination he left academic physics and settled in rural England, supporting his family translating Russian scientific journals. Thirty-eight years later, still living in the same house, he has worked out enough answers to rise from obscurity and capture the attention of the world's physics community.

In the 1970s Barbour began publishing his ideas in respected but slightly unconventional journals, like *The British Journal for the Philosophy of Science* and *Proceedings of the Royal Society A*. He continues to issue papers, most recently with his collaborator Edward Anderson of the University of Cambridge. Barbour's arguments are complex, but his core idea remains simplicity itself: there is no time. "If you try to get your hands on time, it's always slipping through your fingers," Barbour says with his disarming English charm. "My feeling is that people can't get hold of time because it isn't there at all."

Isaac Newton thought of time as a river flowing at the same rate everywhere. Albert Einstein unified space and time into a single entity, but he still held on to the concept of time as a measure of change. In Barbour's view there is no invisible river of time. Instead, he thinks that change merely creates an illusion of time, with each individual moment existing in its own right, complete and whole. He calls these moments "Nows."

"As we live, we seem to move through a succession of Nows. The question is, what are they?" Barbour asks. His answer: Each Now is an arrangement of everything in the universe. "We have the strong impression that things have definite positions relative to each other. I aim to abstract away everything we cannot see, directly or indirectly, and simply keep this idea of many different things coexisting at once. There are simply the Nows, nothing more and nothing less."

Barbour's Nows can be imagined as pages of a novel ripped from the book's spine and tossed randomly onto the floor. Each page is a separate entity. Arranging the pages in some special order and moving through them step by step makes it seem that a story is unfolding. Even so, no matter how we arrange the sheets, each page is complete and independent. For Barbour, reality is just the physics of these Nows taken together as a whole.

"What really intrigues me is that the totality of all possible Nows has a very special structure," he says. "You can think of it as a landscape or country. Each point in this country is a Now, and I call the country Platonia," in reference to Plato's conception of a deeper reality, "because it is timeless and created by perfect mathematical rules. Platonia is the true arena of the universe."

In Platonia all possible configurations of the universe, every possible location of every atom, exist simultaneously. There is no past moment that flows into a future moment; the question of what came before the Big Bang never arises because Barbour's cosmology has no time. The Big Bang is not an event in the distant past; it is just one special place in Platonia.

Our illusion of the past comes because each Now in Platonia contains objects that appear as "records," in Barbour's language. "The only evidence you have of last week is your memory — but memory comes from a stable structure of neurons in your brain now. The only evidence we have of the earth's past are rocks and fossils — but these are just stable structures in the form of an arrangement of minerals we examine in the present. All we have are these records, and we only have them in this Now," Barbour says. In his theory, some Nows are linked to others in Platonia's landscape even though they all exist simultaneously. Those links create the appearance of a sequence from past to future, but there is no actual flow of time from one Now to another.

"Think of the integers," Barbour says. "Every integer exists simultaneously. But some of the integers are linked in structure, like the set of all primes or the numbers you get from the Fibonacci series." Yet the number 3 does not occur in the past of the number 5 any more than the Big Bang exists in the past of the year 2008.

These ideas might sound like the stuff of late-night dorm-room conversations, but Barbour has spent four decades hammering them out in the hard language of mathematical physics. He has blended Platonia with the equations of quantum mechanics to devise a mathematical description of a "changeless" physics. With Irish collaborator Niall Ó Murchadha of the National University of Ireland in Cork, Barbour is continuing to reformulate a time-free version of Einstein's theory.

So What Really Happened?

For each of the alternatives to the Big Bang, it is easier to demonstrate the appeal of the idea than to prove that it is correct. Steinhardt and Turok's cyclic cosmology can account for critical pieces of evidence usually cited to support the Big Bang, but the experiments that could put it over the top are decades away. Carroll's model of the multiverse depends on a speculative interpretation of inflationary cosmology, which is itself only loosely verified.

Barbour stands at the farthest extreme. He has no way to test his concept of Platonia. The power of his ideas rests heavily on the beauty of their formulation and on their capacity to unify physics. "What we are working out now is simple and coherent," Barbour says, "and because of that I believe it is showing us something fundamental."

The payoff that Barbour offers is not just a mathematical solution but a philosophical one. In place of all the conflicting notions about the Big Bang and what came before, he offers a way out. He proposes letting go of the past — of the whole idea of the past — and living fully, happily, in the Now.

ATUL GAWANDE

The Itch

FROM *The New Yorker*

IT WAS STILL SHOCKING to M. how much a few wrong turns could change your life. She had graduated from Boston College with a degree in psychology, married at twenty-five, and had two children, a son and a daughter. She and her family settled in a town on Massachusetts' southern shore. She worked for thirteen years in health care, becoming the director of a residence program for men who'd suffered severe head injuries. But she and her husband began fighting. There were betrayals. By the time she was thirty-two, her marriage had disintegrated. In the divorce, she lost possession of their home, and, amid her financial and psychological struggles, she saw that she was losing her children, too. Within a few years, she was drinking. She began dating someone, and they drank together. After a while, he brought some drugs home, and she tried them. The drugs got harder. Eventually, they were doing heroin, which turned out to be readily available from a street dealer a block away from her apartment.

One day she went to see a doctor because she wasn't feeling well and learned that she had contracted HIV from a contaminated needle. She had to leave her job. She lost visiting rights with her children. And she developed complications from the HIV, including shingles, which caused painful, blistering sores across her scalp and forehead. With treatment, though, her HIV was brought under control. At thirty-six, she entered rehab, dropped the boyfriend, and kicked the drugs. She had two good, quiet years in which she began rebuilding her life. Then she got the itch.

It was right after a shingles episode. The blisters and the pain responded, as they usually did, to acyclovir, an antiviral medication. But this time the area of the scalp that was involved became numb, and the pain was replaced by a constant, relentless itch. She felt it mainly on the right side of her head. It crawled along her scalp, and no matter how much she scratched, it would not go away. "I felt like my inner self, like my brain itself, was itching," she says. And it took over her life just as she was starting to get it back.

Her internist didn't know what to make of the problem. Itching is an extraordinarily common symptom. All kinds of dermatological conditions can cause it: allergic reactions, bacterial or fungal infections, skin cancer, psoriasis, dandruff, scabies, lice, poison ivy, sun damage, or just dry skin. Creams and makeup can cause itch, too. But M. used ordinary shampoo and soap, no creams. And when the doctor examined M.'s scalp she discovered nothing abnormal — no rash, no redness, no scaling, no thickening, no fungus, no parasites. All she saw was scratch marks

The internist prescribed a medicated cream, but it didn't help. The urge to scratch was unceasing and irresistible. "I would try to control it during the day, when I was aware of the itch, but it was really hard," M. said. "At night it was the worst. I guess I would scratch when I was asleep, because in the morning there would be blood on my pillowcase." She began to lose her hair over the itchy area. She returned to her internist again and again. "I just kept haunting her and calling her," M. said. But nothing the internist tried worked, and she began to suspect that the itch had nothing to do with M.'s skin.

Plenty of nonskin conditions can cause itching. Dr. Jeffrey Bernhard, a dermatologist with the University of Massachusetts Medical School, is among the few doctors to study itching systematically (he published the definitive textbook on the subject), and he told me of cases caused by hyperthyroidism, iron deficiency, liver disease, and cancers like Hodgkin's lymphoma. Sometimes the syndrome is very specific. Persistent outer-arm itching that worsens in sunlight is known as brachioradial pruritus, and it's caused by a crimped nerve in the neck. Aquagenic pruritus is recurrent, intense, diffuse itching upon getting out of a bath or shower, and although no one knows the mechanism, it's a symptom of polycythemia vera, a rare condition in which the body produces too many red blood cells.

But M.'s itch was confined to the right side of her scalp. Her viral

count showed that the HIV was quiescent. Additional blood tests
and X-rays were normal. So the internist concluded that M.'s problem was probably psychiatric. All sorts of psychiatric conditions can
cause itching. Patients with psychosis can have cutaneous delusions
— a belief that their skin is infested with, say, parasites, or crawling
ants, or laced with tiny bits of fiberglass. Severe stress and other
emotional experiences can also give rise to a physical symptom like
itching — whether from the body's release of endorphins (natural
opioids, which, like morphine, can cause itching), increased skin
temperature, nervous scratching, or increased sweating. In M.'s
case, the internist suspected trichotillomania, an obsessive-compulsive disorder (OCD) in which patients have an irresistible urge to
pull out their hair.

M. was willing to consider such possibilities. Her life had been
a mess, after all. But the antidepressant medications often prescribed for OCD made no difference. And she didn't actually feel a
compulsion to pull out her hair. She simply felt itchy, on the area of
her scalp that was left numb from the shingles. Although she could
sometimes distract herself from it — by watching television or talking with a friend — the itch did not fluctuate with her mood or
level of stress. The only thing that came close to offering relief was
to scratch.

"Scratching is one of the sweetest gratifications of nature, and as
ready at hand as any," Montaigne wrote. "But repentance follows
too annoyingly close at its heels." For M., certainly, it did: the itching was so torturous, and the area so numb, that her scratching began to go through the skin. At a later office visit, her doctor found a
silver-dollar-size patch of scalp where skin had been replaced by
scab. M. tried bandaging her head, wearing caps to bed. But her
fingernails would always find a way to her flesh, especially while she
slept.

One morning, after she was awakened by her bedside alarm, she
sat up and, she recalled, "this fluid came down my face, this greenish liquid." She pressed a square of gauze to her head and went to
see her doctor again. M. showed the doctor the fluid on the dressing. The doctor looked closely at the wound. She shined a light on
it and in M.'s eyes. Then she walked out of the room and called an
ambulance. Only in the Emergency Department at Massachusetts
General Hospital, after the doctors started swarming, and one told

her she needed surgery now, did M. learn what had happened. She had scratched through her skull during the night — and all the way into her brain.

Itching is a most peculiar and diabolical sensation. The definition offered by the German physician Samuel Hafenreffer in 1660 has yet to be improved upon: an unpleasant sensation that provokes the desire to scratch. Itch has been ranked, by scientific and artistic observers alike, among the most distressing physical sensations one can experience. In Dante's *Inferno,* falsifiers were punished by "the burning rage/of fierce itching that nothing could relieve":

> The way their nails scraped down upon the scabs
> Was like a knife scraping off scales from carp . . .
>
> "O you there tearing at your mail of scabs
> And even turning your fingers into pincers,"
> My guide began addressing one of them,
>
> "Tell us are there Italians among the souls
> Down in this hole and I'll pray that your nails
> Will last you in this task eternally."

Though scratching can provide momentary relief, it often makes the itching worse. Dermatologists call this the itch-scratch cycle. Scientists believe that itch, and the accompanying scratch reflex, evolved in order to protect us from insects and clinging plant toxins — from such dangers as malaria, yellow fever, and dengue, transmitted by mosquitoes; from tularemia, river blindness, and sleeping sickness, transmitted by flies; from typhus-bearing lice, plague-bearing fleas, and poisonous spiders. The theory goes a long way toward explaining why itch is so exquisitely tuned. You can spend all day without noticing the feel of your shirt collar on your neck, and yet a single stray thread poking out, or a louse's fine legs brushing by, can set you scratching furiously.

But how, exactly, itch works has been a puzzle. For most of medical history, scientists thought that itching was merely a weak form of pain. Then, in 1987, the German researcher H. O. Handwerker and his colleagues used mild electric pulses to drive histamine, an itch-producing substance that the body releases during allergic reactions, into the skin of volunteers. As the researchers increased the dose of histamine, they found that they were able to increase the intensity of itch the volunteers reported, from the barely appreciable to the "maximum imaginable." Yet the volunteers never felt an increase in pain. The scientists concluded that itch and pain are entirely separate sensations, transmitted along different pathways.

Despite centuries spent mapping the body's nervous circuitry, scientists had never noticed a nerve specific for itch. But now the hunt was on, and a group of Swedish and German researchers embarked upon a series of tricky experiments. They inserted ultrathin metal electrodes into the skin of paid volunteers and wiggled them around until they picked up electrical signals from a single nerve fiber. Computers subtracted the noise from other nerve fibers crossing through the region. The researchers would then spend hours — as long as the volunteer could tolerate it — testing different stimuli on the skin in the area (a heated probe, for example, or a fine paintbrush) to see what would get the nerve to fire and what the person experienced when it did.

They worked their way through fifty-three volunteers. Mostly, they encountered well-known types of nerve fibers that respond to temperature or light touch or mechanical pressure. "That feels warm," a volunteer might say, or "That feels soft," or "Ouch! Hey!" Several times the scientists came across a nerve fiber that didn't respond to any of these stimuli. When they introduced a tiny dose of histamine into the skin, however, they observed a sharp electrical response in some of these nerve fibers, and the volunteer would experience an itch. They announced their discovery in a 1997 paper: they'd found a type of nerve that was specific for itch.

Unlike, say, the nerve fibers for pain, each of which covers a millimeter-size territory, a single itch fiber can pick up an itchy sensation more than three inches away. The fibers also turned out to have extraordinarily low conduction speeds, which explained why itchiness is so slow to build and so slow to subside.

Other researchers traced these fibers to the spinal cord and all the way to the brain. Examining functional PET-scan studies in healthy human subjects who had been given mosquito-bite-like histamine injections, they found a distinct signature of itch activity. Several specific areas of the brain light up: the part of the cortex that tells you where on your body the sensation occurs; the region that governs your emotional responses, reflecting the disagreeable nature of itch; and the limbic and motor areas that process irresistible urges (such as the urge to use drugs, among the addicted, or to overeat, among the obese), reflecting the ferocious impulse to scratch.

Now various phenomena became clear. Itch, it turns out, is indeed inseparable from the desire to scratch. It can be triggered chemically (by the saliva injected when a mosquito bites, say) or mechanically (from the mosquito's legs, even before it bites). The itch-scratch reflex activates higher levels of your brain than the spinal-cord-level reflex that makes you pull your hand away from a flame. Brain scans also show that scratching diminishes activity in brain areas associated with unpleasant sensations.

But some basic features of itch remained unexplained — features that make itch a uniquely revealing case study. On the one hand, our bodies are studded with receptors for itch, as they are with receptors for touch, pain, and other sensations; this provides an alarm system for harm and allows us to safely navigate the world. But why does a feather brushed across the skin sometimes itch and at other times tickle? (Tickling has a social component: you can make yourself itch, but only another person can tickle you.) And, even more puzzling, how is it that you can make yourself itchy just by thinking about it?

Contemplating what it's like to hold your finger in a flame won't make your finger hurt. But simply writing about a tick crawling up the nape of one's neck is enough to start my neck itching. Then my scalp. And then this one little spot along my flank where I'm beginning to wonder whether I should check to see if there might be something there. In one study, a German professor of psychosomatics gave a lecture that included, in the first half, a series of what might be called itchy slides, showing fleas, lice, people scratching, and the like, and, in the second half, more benign slides, with pictures of soft down, baby skin, bathers. Video cameras recorded the audience. Sure enough, the frequency of scratching among people in the audience increased markedly during the first half and decreased during the second. Thoughts made them itch.

We now have the nerve map for itching, as we do for other sensations. But a deeper puzzle remains: how much of our sensations and experiences do nerves really explain?

*

In the operating room, a neurosurgeon washed out and debrided M.'s wound, which had become infected. Later a plastic surgeon covered it with a graft of skin from her thigh. Though her head was wrapped in layers of gauze and she did all she could to resist the still furious itchiness, she awoke one morning to find that she had rubbed the graft away. The doctors returned her to the operating room for a second skin graft, and this time they wrapped her hands as well. She rubbed the graft away again anyway.

"They kept telling me I had OCD," M. said. A psychiatric team was sent in to see her each day, and the resident would ask her, "As a child, when you walked down the street did you count the lines? Did you do anything repetitive? Did you have to count everything you saw?" She kept telling him no, but he seemed skeptical. He tracked down her family and asked them, but they said no, too. Psychology tests likewise ruled out obsessive-compulsive disorder. They showed depression, though, and, of course, there was the history of addiction. So the doctors still thought her scratching was from a psychiatric disorder. They gave her drugs that made her feel logy and sleep a lot. But the itching was as bad as ever, and she still woke up scratching at that terrible wound.

One morning she found, as she put it, "this very bright and happy-looking woman standing by my bed. She said, 'I'm Dr. Oaklander,'" M. recalled. "I thought, Oh great. Here we go again. But she explained that she was a neurologist, and she said, 'The first thing I want to say to you is that I don't think you're crazy. I don't think you have OCD.' At that moment, I really saw her grow wings and a halo," M. told me. "I said, 'Are you sure?' And she said, 'Yes. I have heard of this before.'"

Anne Louise Oaklander was about the same age as M. Her mother is a prominent neurologist at Albert Einstein College of Medicine in New York, and she'd followed her mother into the field. Oaklander had specialized in disorders of peripheral nerve sensation — disorders like shingles. Although pain is the most common symptom of shingles, Oaklander had noticed during her training that some patients also had itching, occasionally severe, and seeing M. reminded her of one of her shingles patients. "I remember standing in a hallway talking to her, and what she complained about — her major concern — was that she was tormented by this terrible itch over the eye where she had had shingles," she told me. When Oaklander looked at M., she thought that something wasn't right. It took a moment to realize why. "The itch was so severe, she had scratched off her eyebrow."

Oaklander tested the skin near M.'s wound. It was numb to temperature, touch, and pinprick. Nonetheless, it was itchy, and when it was scratched or rubbed M. felt the itchiness temporarily subside. Oaklander injected a few drops of local anesthetic into the skin. To M.'s surprise, the itching stopped — instantly and almost entirely. This was the first real relief she'd had in more than a year.

It was an imperfect treatment, though. The itch came back when the anesthetic wore off, and, although Oaklander tried having M. wear an anesthetic patch over the wound, the effect diminished over time. Oaklander did not have an explanation for any of this. When she took a biopsy of the itchy skin, it showed that 96 percent of the nerve fibers were gone. So why was the itch so intense?

Oaklander came up with two theories. The first was that those few remaining nerve fibers were itch fibers and, with no other fibers around to offer competing signals, they had become constantly active. The second theory was the opposite. The nerves were dead, but perhaps the itch system in M.'s brain had gone haywire, running on a loop all its own.

The second theory seemed less likely. If the nerves to her scalp were dead, how would you explain the relief she got from scratching or from the local anesthetic? Indeed, how could you explain the itch in the first place? An itch without nerve endings didn't make sense. The neurosurgeons stuck with the first theory; they offered to cut the main sensory nerve to the front of M.'s scalp and abolish the itching permanently. Oaklander, however, thought that the second theory was the right one — that this was a brain problem, not a nerve problem — and that cutting the nerve would do more harm than good. She argued with the neurosurgeons, and she advised M. not to let them do any cutting.

"But I was desperate," M. told me. She let them operate on her, slicing the supraorbital nerve above the right eye. When she woke up, a whole section of her forehead was numb — and the itching was gone. A few weeks later, however, it came back, in an even wider expanse than before. The doctors tried pain medications, more psychiatric medications, more local anesthetic. But the only thing that kept M. from tearing her skin and skull open again, the doctors found, was to put a foam football helmet on her head and bind her wrists to the bedrails at night.

She spent the next two years committed to a locked medical ward in a rehabilitation hospital — because, although she was not mentally ill, she was considered a danger to herself. Eventually, the staff worked out a solution that did not require binding her to the bedrails. Along with the football helmet, she had to wear white mitts that were secured around her wrists by surgical tape. "Every bedtime, it looked like they were dressing me up for Halloween — me and the guy next to me," she told me.

"The guy next to you?" I asked. He had had shingles on his neck, she explained, and also developed a persistent itch. "Every night they would wrap up his hands and wrap up mine." She spoke more softly now. "But I heard he ended up dying from it, because he scratched into his carotid artery."

I met M. seven years after she'd been discharged from the rehabilitation hospital. She is forty-eight now. She lives in a three-room apartment, with a crucifix and a bust of Jesus on the wall and the low yellow light of table lamps strung with beads over their shades. Stacked in a wicker basket next to her coffee table were Rick Warren's *Purpose Driven Life, People,* and the latest issue of *Neurology Now,* a magazine for patients. Together, they summed up her struggles, for she is still fighting the meaninglessness, the isolation, and the physiology of her predicament.

She met me at the door in a wheelchair; the injury to her brain had left her partially paralyzed on the left side of her body. She remains estranged from her children. She has not, however, relapsed into drinking or drugs. Her HIV remains under control. Although the itch on her scalp and forehead persists, she has gradually learned to protect herself. She trims her nails short. She finds ways to distract herself. If she must scratch, she tries to rub gently instead. And if that isn't enough, she uses a soft toothbrush or a rolled-up terry cloth. "I don't use anything sharp," she said. The two years that she spent bound up in the hospital seemed to have broken the nighttime scratching. At home she found that she didn't need to wear the helmet and gloves anymore.

Still, the itching remains a daily torment. "I don't normally tell people this," she said, "but I have a fantasy of shaving off my eyebrow and taking a metal-wire grill brush and scratching away."

Some of her doctors have not been willing to let go of the idea that this has been a nerve problem all along. A local neurosurgeon told her that the original operation to cut the sensory nerve to her scalp must not have gone deep enough. "He wants to go in again," she told me.

A new scientific understanding of perception has emerged in the past few decades, and it has overturned classical, centuries-long beliefs about how our brains work — though it has apparently not penetrated the medical world yet. The old understanding of perception is what neuroscientists call "the naive view," and it is the view that most people, in or out of medicine, still have. We're inclined to think that people normally perceive things in the world directly. We believe that the hardness of a rock, the coldness of an ice cube, the itchiness of a sweater, are picked up by our nerve endings, transmitted through the spinal cord like a message through a wire, and decoded by the brain.

In his 1710 *Treatise Concerning the Principles of Human Knowledge*, the Irish philosopher George Berkeley objected to this view. We do not know the world of objects, he argued; we know only our mental ideas of objects. "Light and colours, heat and cold, extension and figures — in a word, the things we see and feel — what are they but so many sensations, notions, ideas?" Indeed, he concluded, the objects of the world are likely just inventions of the mind, put in there by God. To which Samuel Johnson famously responded by kicking a large stone and declaring, "I refute it thus!"

Still, Berkeley had recognized some serious flaws in the direct-perception theory — in the notion that when we see, hear, or feel we are just taking in the sights, sounds, and textures of the world. For one thing, it cannot explain how we experience things that seem physically real but aren't: sensations of itching that arise from nothing more than itchy thoughts; dreams that can seem indistinguishable from reality; phantom sensations that amputees have in their missing limbs. And the more we examine the actual nerve transmissions we receive from the world outside, the more inadequate they seem.

Our assumption had been that the sensory data we receive from our eyes, ears, nose, fingers, and so on contain all the information we need for perception and that perception must work something like a radio. It's hard to conceive that a Boston Symphony Orchestra concert is in a radio wave. But it is. So you might think that it's the same with the signals we receive — that if you hooked up someone's nerves to a monitor you could watch what the person is experiencing as if it were a television show.

Yet as scientists set about analyzing the signals, they found them to be radically impoverished. Suppose someone is viewing a tree in a clearing. Given simply the transmissions along the optic nerve from the light entering the eye, one would not be able to reconstruct the three-dimensionality, or the distance, or the detail of the bark — attributes that we perceive instantly.

Or consider what neuroscientists call "the binding problem." Tracking a dog as it runs behind a picket fence, all that your eyes receive is separated vertical images of the dog, with large slices missing. Yet somehow you perceive the mutt to be whole, an intact entity traveling through space. Put two dogs together behind the fence and you don't think they've morphed into one. Your mind now configures the slices as two independent creatures.

The images in our mind are extraordinarily rich. We can tell if something is liquid or solid, heavy or light, dead or alive. But the information we work from is poor — a distorted, two-dimensional transmission with entire spots missing. So the mind fills in most of the picture. You can get a sense of this from brain-anatomy studies. If visual sensations were primarily received rather than constructed by the brain, you'd expect that most of the fibers going to the brain's primary visual cortex would come from the retina. Instead, scientists have found that only 20 percent do; 80 percent come downward from regions of the brain governing functions like memory. Richard Gregory, a prominent British neuropsychologist, estimates that visual perception is more than 90 percent memory and less than 10 percent sensory nerve signals. When Oaklander theorized that M.'s itch was endogenous rather than generated by peripheral nerve signals, she was onto something important.

The fallacy of reducing perception to reception is especially clear when it comes to phantom limbs. Doctors have often explained such sensations as a matter of inflamed or frayed nerve endings in the stump sending aberrant signals to the brain. But this explanation should long ago have been suspect. Efforts by surgeons to cut back on the nerve typically produce same results that M. had when they cut the sensory nerve to her forehead: a brief period of relief followed by a return of the sensation.

Moreover, the feelings people experience in their phantom limbs are far too varied and rich to be explained by the random firings of a bruised nerve. People report not just pain but also sensations of sweatiness, heat, texture, and movement in a missing limb. There is no experience people have with real limbs that they do not experience with phantom limbs. They feel their phantom leg swinging, water trickling down a phantom arm, a phantom ring becoming too tight for a phantom digit. Children have used phantom fingers to count and solve arithmetic problems. V. S. Ramachandran, an eminent neuroscientist at the University of California, San Diego, has written up the case of a woman who was born with only stumps at her shoulders, and yet, as far back as she could remember, felt herself to have arms and hands; she even feels herself gesticulating when she speaks. And phantoms do not occur just in limbs. Around half of women who have undergone a mastectomy experience a phantom breast, with the nipple being the most vivid part. You've likely had an experience of phantom sensation yourself. When the dentist gives you a local anesthetic, and your lip goes numb, the nerves go dead. Yet you don't feel your lip disappear. Quite the opposite: it feels larger and plumper than normal, even though you can see in a mirror that the size hasn't changed.

The account of perception that's starting to emerge is what we might call the "brain's best guess" theory: perception is the brain's best guess about what is happening in the outside world. The mind integrates scattered, weak, rudimentary signals from a variety of sensory channels, information from past experiences, and hardwired processes, and produces a sensory experience full of brain-provided color, sound, texture, and meaning. We see a friendly yellow Labrador bounding behind a picket fence not because that is the transmission we receive but because this is the perception our weaver-brain assembles as its best hypothesis of what is out there from the slivers of information we get. Perception is inference.

The theory — and a theory is all it is right now — has begun to make sense of some bewildering phenomena. Among them is an experiment that Ramachandran performed with volunteers who had phantom pain in an amputated arm. They put their surviving arm through a hole in the side of a box with a mirror inside, so that, peering through the open top, they would see their arm and its mirror image, as if they had two arms. Ramachandran then asked them to move both their intact arm and, in their mind, their phantom arm — to pretend that they were conducting an orchestra, say. The patients had the sense that they had two arms again. Even though they knew it was an illusion, it provided immediate relief. People who for years had been unable to unclench their phantom fist suddenly felt their hand open; phantom arms in painfully contorted positions could relax. With daily use of the mirror box over weeks, patients sensed their phantom limbs actually shrink into their stumps and, in several instances, completely vanish. Researchers at Walter Reed Army Medical Center recently published the results of a randomized trial of mirror therapy for soldiers with phantom-limb pain, showing dramatic success.

A lot about this phenomenon remains murky, but here's what the new theory suggests is going on: when your arm is amputated, nerve transmissions are shut off, and the brain's best guess often seems to be that the arm is still there, but paralyzed, or clenched, or beginning to cramp up. Things can stay like this for years. The mirror box, however, provides the brain with new visual input — however illusory — suggesting motion in the absent arm. The brain has to incorporate the new information into its sensory map of what's happening. Therefore, it guesses again, and the pain goes away.

The new theory may also explain what was going on with M.'s itch. The shingles destroyed most of the nerves in her scalp. And for whatever reason, her brain surmised from what little input it had that something horribly itchy was going on — that perhaps a whole army of ants were crawling back and forth over just that patch of skin. There wasn't any such thing, of course. But M.'s brain has received no contrary signals that would shift its assumptions. So she itches.

Not long ago, I met a man who made me wonder whether such phantom sensations are more common than we realize. H. was forty-eight, in good health, an officer at a Boston financial services company living with his wife in a western suburb, when he made passing mention of an odd pain to his internist. For at least twenty years, he said, he'd had a mild tingling running along his left arm and down the left side of his body, and, if he tilted his neck forward at a particular angle, it became a pronounced electrical jolt. The internist recognized this as Lhermitte's sign, a classic symptom that can indicate multiple sclerosis, Vitamin B_{12} deficiency, or spinal-cord compression from a tumor or a herniated disk. An MRI revealed a cavernous hemangioma, a pea-size mass of dilated blood vessels, pressing into the spinal cord in his neck. A week later, while the doctors were still contemplating what to do, it ruptured.

"I was raking leaves out in the yard and, all of a sudden, there was an explosion of pain and my left arm wasn't responding to my brain," H. said when I visited him at home. Once the swelling subsided, a neurosurgeon performed a tricky operation to remove the tumor from the spinal cord. The operation was successful, but afterward H. began experiencing a constellation of strange sensations. His left hand felt cartoonishly large — at least twice its actual size. He developed a constant burning pain along an inch-wide ribbon extending from the left side of his neck all the way down his arm. And an itch crept up and down along the same band, which no amount of scratching would relieve.

H. has not accepted that these sensations are here to stay — the prospect is too depressing — but they've persisted for eleven years now. Although the burning is often tolerable during the day, the slightest thing can trigger an excruciating flare-up — a cool breeze across the skin, the brush of a shirtsleeve or a bed sheet. "Sometimes I feel that my skin has been flayed and my flesh is exposed, and any touch is just very painful," he told me. "Sometimes I feel that there's an ice pick or a wasp sting. Sometimes I feel that I've been splattered with hot cooking oil."

For all that, the itch has been harder to endure. H. has developed calluses from the incessant scratching. "I find I am choosing itch relief over the pain that I am provoking by satisfying the itch," he said.

He has tried all sorts of treatments — medications, acupuncture, herbal remedies, lidocaine injections, electrical-stimulation therapy. But nothing has really worked, and the condition forced him to retire in 2001. He now avoids leaving the house. He gives himself projects. Last year he built a three-foot stone wall around his yard, slowly placing the stones by hand. But he spends much of his day, after his wife has left for work, alone in the house with their three cats, his shirt off and the heat turned up, trying to prevent a flare-up.

His neurologist introduced him to me, with his permission, as an example of someone with severe itching from a central rather than a peripheral cause. So one morning we sat in his living room trying to puzzle out what was going on. The sun streamed in through a big bay window. One of his cats, a scraggly brown tabby, curled up beside me on the couch. H. sat in an armchair in a baggy purple T-shirt he'd put on for my visit. He told me that he thought his problem was basically a "bad switch" in his neck where the tumor had been, a kind of loose wire sending false signals to his brain. But I told him about the increasing evidence that our sensory experiences are not sent to the brain but originate in it. When I got to the example of phantom-limb sensations, he perked up. The experiences of phantom-limb patients sounded familiar to him. When I mentioned that he might want to try the mirror-box treatment, he agreed. "I have a mirror upstairs," he said.

He brought a cheval glass down to the living room, and I had him stand with his chest against the side of it, so that his troublesome left arm was behind it and his normal right arm was in front. He tipped his head so that when he looked into the mirror the image of his right arm seemed to occupy the same position as his left arm. Then I had him wave his arms, his actual arms, as if he were conducting an orchestra.

The first thing he expressed was disappointment. "It isn't quite like looking at my left hand," he said. But then suddenly it was.

"Wow!" he said. "Now, this is odd."

After a moment or two, I noticed that he had stopped moving his left arm. Yet he reported that he still felt as if it were moving. What's more, the sensations in it had changed dramatically. For the first time in eleven years, he felt his left hand "snap" back to normal size. He felt the burning pain in his arm diminish. And the itch, too, was dulled.

"This is positively bizarre," he said.

He still felt the pain and the itch in his neck and shoulder, where the image in the mirror cut off. And when he came away from the mirror, the aberrant sensations in his left arm returned. He began using the mirror a few times a day, for fifteen minutes or so at a stretch, and I checked in with him periodically.

"What's most dramatic is the change in the size of my hand," he says. After a couple of weeks, his hand returned to feeling normal in size all day long.

The mirror also provided the first effective treatment he has had for the flares of itch and pain that sporadically seize him. Where once he could do nothing but sit and wait for the torment to subside — it sometimes took an hour or more — he now just pulls out the mirror. "I've never had anything like this before," he said. "It's my magic mirror."

There have been other, isolated successes with mirror treatment. In Bath, England, several patients suffering from what is called complex regional pain syndrome — severe, disabling limb sensations of unknown cause — were reported to have experienced complete resolution after six weeks of mirror therapy. In California, mirror therapy helped stroke patients recover from a condition known as hemineglect, which produces something like the opposite of a phantom limb — these patients have a part of the body they no longer realize is theirs.

Such findings open up a fascinating prospect: perhaps many patients whom doctors treat as having a nerve injury or a disease have, instead, what might be called sensor syndromes. When your car's dashboard warning light keeps telling you that there is an engine failure, but the mechanics can't find anything wrong, the sensor itself may be the problem. This is no less true for human beings. Our sensations of pain, itch, nausea, and fatigue are normally protective. Unmoored from physical reality, however, they can become a nightmare: M., with her intractable itching, and H., with his constellation of strange symptoms — but perhaps also the hundreds of thousands of people in the United States alone who suffer from conditions like chronic back pain, fibromyalgia, chronic pelvic pain, tinnitus, temporomandibular joint disorder, or repetitive strain injury, for whom, typically, no amount of imaging, nerve testing, or surgery manages to uncover an anatomical explanation. Doctors have persisted in treating these conditions as nerve or tissue problems — engine failures, as it were. We get under the hood and remove this, replace that, snip some wires. Yet still the sensor keeps going off.

So we get frustrated. "There's nothing wrong," we'll insist. And the next thing you know, we're treating the driver instead of the problem. We prescribe tranquilizers, antidepressants, escalating doses of narcotics. And the drugs often do make it easier for people to ignore the sensors, even if they are wired right into the brain. The mirror treatment, by contrast, targets the deranged sensor system itself. It essentially takes a misfiring sensor — a warning system functioning under an illusion that something is terribly wrong out in the world it monitors — and feeds it an alternate set of signals that calm it down. The new signals may even reset the sensor.

This may help explain, for example, the success of the advice that back specialists now commonly give. Work through the pain, they tell many of their patients, and, surprisingly often, the pain goes away. It had been a mystifying phenomenon. But the picture now seems clearer. Most chronic back pain starts as acute back pain — after a fall, say. Usually the pain subsides as the injury heals. But in some cases the pain sensors continue to light up long after the tissue damage is gone. In such instances, working through the pain may offer the brain contradictory feedback — a signal that ordinary activity does not, in fact, cause physical harm. And so the sensor resets.

This understanding of sensation points to an entire new array of potential treatments — based not on drugs or surgery but, instead, on the careful manipulation of our perceptions. Researchers at the University of Manchester, in England, have gone a step beyond mirrors and fashioned an immersive virtual-reality system for treating patients with phantom-limb pain. Detectors transpose movement of real limbs into a virtual world where patients feel they are actually moving, stretching, even playing a ball game. So far, five patients have tried the system, and they have all experienced a reduction in pain. Whether those results will last has yet to be established. But the approach raises the possibility of designing similar systems to help patients with other sensor syndromes. How, one wonders, would someone with chronic back pain fare in a virtual world? The Manchester study suggests that there may be many ways to fight our phantoms.

I called Ramachandran to ask him about M.'s terrible itch. The sensation may be a phantom, but it's on her scalp, not in a limb, so it seemed unlikely that his mirror approach could do anything for her. He told me about an experiment in which he put ice-cold water in people's ears. This confuses the brain's position sensors, tricking subjects into thinking that their heads are moving, and in certain phantom-limb and stroke patients the illusion corrected their misperceptions, at least temporarily. Maybe this would help M., he said. He had another idea. If you take two mirrors and put them at right angles to each other, you will get a nonreversed mirror image. Looking in, the right half of your face appears on the left and the left half appears on the right. But unless you move, he said, your brain may not realize that the image is flipped.

"Now, suppose she looks in this mirror and scratches the left side of her head. No, wait — I'm thinking out loud here — suppose she looks and you have someone else touch the left side of her head. It'll look — maybe it'll feel — like you're touching the right side of her head." He let out an impish giggle. "Maybe this would make her itchy right scalp feel more normal." Maybe it would encourage her brain to make a different perceptual inference; maybe it would press reset. "Who knows?" he said.

It seemed worth a try.

DAVID GRIMM

The Mushroom Cloud's Silver Lining

FROM *Science*

THE TWO MUMMIFIED BODIES in the Vienna apartment told a sad tale. The reclusive elderly sisters had clearly been dead for several years, but no one had noticed; neighbors in the upper-middle-class complex believed they had merely moved away. Stale bank accounts finally tipped off the police, who discovered the remains in December 1992.

Investigators found no evidence of foul play, so they focused on the question of who died first. Both sisters had large pensions and separate life insurance policies, and the insurance company of the woman who died last would collect the bulk of the funds. "There was a lot of money at stake," says Walter Kutschera, a physicist at the Vienna Environmental Research Accelerator at the University of Vienna in Austria. Not long after the bodies were found, a scientist from the university's forensics department approached Kutschera and his colleague, Eva Maria Wild, to ask if they could help crack the case. The forensics expert knew the pair had been using radiocarbon dating to determine the age of archaeological samples, and he wondered if the same technique could reveal the year each sister had died.

It couldn't. Radiocarbon dating is a blunt instrument that relies on the slow decay of a form of carbon known as carbon 14 (^{14}C), which is incorporated into animals during their lifetime. The method works well for samples that are tens of thousands of years old, but it's accurate only to within a few hundred years.

Wild and Kutschera had another idea. Aboveground testing of nuclear weapons after World War II had injected ^{14}C into Earth's atmosphere, creating an abnormally high level of the isotope, which has been tapering off since then. If the researchers could measure the amount of ^{14}C in something carbon-based that the sisters had generated just before death — fats in the bone, for example — and compare it with historic levels of ^{14}C in the atmosphere, they should be able to tell in what year each sister had expired.

It worked. Wild and Kutschera found that one sister had died in 1988 and the other in 1989. "One sister lived for some time next to the dead one," says Wild. Investigators closed the case, and Wild and Kutschera returned to dating ancient bones and seeds. But it soon became clear that the "bomb-pulse" technique had much more to offer. In the past decade, thanks largely to the pioneering work of an Australian postdoc with a taste for trying new things, groups have begun using the strategy for diverse causes, such as identifying disaster victims, authenticating wine vintages, and tackling some of the most controversial questions in biology, including whether the human brain generates neurons throughout life.

From Pet Shop to Slaughterhouse

The year 2001 started well for Kirsty Spalding, but by the end of it she would be knee-deep in a failing project. The twenty-nine-year-old had just finished her graduate work in neuroscience at the University of Western Australia in Perth, and she was planning on spending a year in Europe as a postdoc before moving to the United States. On her way to interview at a couple of prospective labs at the Karolinska Institute in Stockholm, Sweden, Spalding caught a talk by Jonas Frisén, a prominent stem-cell researcher there. "It wasn't what I had planned on doing," says Spalding, referring to Frisén's work on the formation of new neurons in the brain — a process called neurogenesis. "But I found him very personable and the work very interesting."

A few months later, Spalding was in Frisén's lab, trying to map neurogenesis in the zebra-fish brain. But neither she nor her labmates had worked with the animal before, and they weren't aware that technical suppliers provided fish specially bred for laboratory study. Instead, Spalding biked over to a local pet shop and brought a few zebra fish back to the lab. Needless to say, the experiments didn't work.

Her mentor didn't lose faith, however. "I could tell that Kirsty liked challenges and that she was extremely entrepreneurial," says Frisén. That made her perfect for a new project he had in mind. Familiar with the bomb-pulse work done by Wild and Kutschera, Frisén wondered if it could be applied to DNA. When a cell divides, ^{14}C in the environment is incorporated in new chromosomes, and thus the DNA effectively takes a snapshot of the amount of atmospheric ^{14}C — and hence the birth date — of the cell. If Frisén could exploit this, he might be able to show whether humans generate new brain cells throughout life — a central question in neuroscience. But no one would take on the project. Postdoc after postdoc turned him down, calling the work too risky and too difficult. When Frisén saw Spalding with the zebra fish, he knew he had found someone who wouldn't be daunted.

Spalding agreed. "I liked the problem-solving aspect of it, and I didn't have the burden of knowledge to know how difficult it would be," she says with a laugh. Spalding's planned one-year sojourn in Europe suddenly became an indefinite commitment.

To address neurogenesis in humans, Spalding needed brains from an animal with a similar life span, so she turned to horses, which can live more than twenty-five years. That meant trips to the local slaughterhouse. "I would watch them walk the horse in . . . and then they would chop off its head and hand it to me," recalls Spalding, who had to excavate the skulls herself. "It's not so easy to hack your way into a horse's head . . . It was not pretty."

Brains in hand, Spalding still had challenges to overcome. She had to measure a scarce isotope, ^{14}C, which makes up only one part per trillion of all of the carbon in the atmosphere. Most ^{14}C comes from cosmic-ray collisions with nitrogen, but when the United States, the former Soviet Union, and other nations detonated more than five hundred nuclear warheads aboveground in the 1950s and '60s, the atmospheric ^{14}C level doubled. It began to dissipate only when the Limited Test Ban Treaty of 1963 moved atomic tests underground.

Despite these elevated atmospheric concentrations, only about one atom of ^{14}C incorporates into every fifteen cells. So relatively huge amounts of tissue — up to 5 grams, depending on the part of the body it comes from — are needed for even the world's most powerful isotope detectors to spot it. Horse brains were big enough to provide that amount, but Spalding also had to find a way to sift through a custard of fat, glia, and fibroblasts for the neurons she needed. After taking nearly a year to develop a technique, she was ready to pin ages on neurons and enter the ongoing fray over neurogenesis.

The Brain War

Pasko Rakic is a five-star general in a conflict that's been raging for more than a decade in the neuroscience field. The Yale University neuroscientist, who did pioneering work in how the primate brain forms, has famously established the beachhead position that the human cerebral cortex — a region key for memory, language, and consciousness — does not make new neurons after development. He's often made the point that adult neurogenesis would be counterproductive, disrupting already formed memories, for example.

But in 1998 a research team found evidence to the contrary. The team gave people with terminal cancer a synthetic compound called bromodeoxyuridine (BrdU), which inserts into newly synthesized DNA and thus serves as a marker for new cells. The compound was supposed to gauge tumor growth, but it also showed up in the hippocampus, the brain's learning and memory center. A year later, the Princeton University neuroscientist Elizabeth Gould bolstered the case for ongoing neurogenesis in the brain by giving adult macaques BrdU and finding it in the neocortex, a region responsible for language and consciousness in humans. But two years after that, Rakic injected a different DNA marker into monkeys and saw no new neurons in the adult brain. The field has been divided ever since.

"It's been extremely difficult to get any information in humans," says Gerd Kempermann, a neurogenesis expert at the Center for Regenerative Therapies Dresden in Germany. BrdU is toxic, so it can't be given to healthy people, and Rakic has expressed concern that the compound confuses cells into dividing, leading to false positives.

But ^{14}C doesn't have that problem. It's not toxic, and, like it or not, we've all absorbed it. "All of humanity is labeled," as Kutschera puts it.

As the salvos continued in the neurogenesis debate, Spalding had proved that she could use the bomb-pulse technique to date brain cells in horses. She shipped her first human samples — from the brain's visual center, the occipital cortex — to Bruce Buchholz, who runs an isotope detector the size of a basketball court at Lawrence Livermore National Laboratory in California. Although nonhuman studies had suggested that the occipital cortex was a hotbed of neurogenesis, the ^{14}C data collected by Buchholz indicated that human neurons from this region had the same birth date as the people they came from. That meant no new visual neurons for adults. A year later, Spalding and colleagues found similar results in the human neocortex.

"It's really extraordinary work, and it's extremely clever," says Kempermann. "I think many people will take it as the final word in the debate." Still, Gould notes that other regions of the human brain — such as the hippocampus — have yet to be tested with the technique. And she says that because the bomb-pulse method doesn't target individual cells, it may not be sensitive enough to pick up a small population of neurons that does divide and that could contribute to repair and learning. Spalding was in the midst of addressing those questions when disaster struck a continent away.

CSI: Sweden

"Total chaos." That's how Stockholm's former chief medical examiner, Henrik Druid, describes the scene as bodies piled up at the Karolinska Institute morgue in the wake of the 2004 Indian Ocean tsunami, which killed more than 200,000 people, including more than 500 Swedish tourists. "The bodies were so badly decomposed, you couldn't tell the teenagers from the old people," he says.

Hoping to help, Spalding approached Druid with some intriguing findings from her days at the slaughterhouse. In addition to analyzing horses' brains, she had looked at their teeth, showing that because enamel is permanent and forms early, its ^{14}C levels give an accurate estimate of the animal's age. Spalding asked Druid if the technique might be useful to him.

"At first I was skeptical," he recalls. But Druid didn't have many options. In the confusion surrounding the disaster, identifying materials, such as X-rays and DNA samples from relatives, had not been shipped with the bodies. "If you have no clue to the identity of a person, age and sex are the most important way to limit the search," he says. Anthropologists are only accurate to within about ten years when trying to determine age from a skeleton. So, aided by Spalding, Druid applied the bomb-pulse technique to the teeth of six tsunami victims. After adding the time it takes for human enamel to form (about twelve years for wisdom teeth, for example), they were able to predict the ages of every victim to within 1.6 years, as borne out by the identifying materials that eventually arrived at Karolinska.

With further refinement, Druid has shaved the accuracy down to one year, and he's now using the approach to help Swedish investigators crack two unsolved homicides. "This is going to be very, very valuable for criminal investigation," says Druid. "In a year or two, you're going to begin seeing cases in the newspaper that were solved with this method." Spalding too has begun working with Swedish police — as well as with investigators in Canada — and she eventually hopes to set up a company to perform the tooth analysis. In preparation, she has taken business classes at night, while forging ahead with her brain work — and a new project that would send her spinning in an entirely different direction.

The Fat Offensive

In 2005 Spalding was presenting her brain findings at Karolinska when a member of the audience approached her. "A Ph.D. student came up to me and said he thought the ^{14}C work was something his dad would be interested in," she says. The father — a prominent researcher at Karolinska named Peter Arner — was grappling with a debate not unlike the one faced by the neurogenesis community.

This time the issue was fat. "If you go to any textbook, it will tell you that once a fat cell is born, you've got it forever," explains John Prins, an expert on fat-cell turnover at the University of Queensland in Brisbane, Australia. But there were some who believed that the blubber on our bellies and hips is constantly dying and being replenished. It's not just an academic debate: if you can make the body destroy more fat than it creates, you've got a ticket to weight loss.

But no one could conclusively address the question. "The tech-

niques we have for measuring fat turnover are insufficiently sensitive and fairly inaccurate," says Prins. The best researchers could
do was have volunteers drink heavy water, which contains elevated
levels of an isotope of hydrogen known as deuterium, and look for
that isotope in fat cells. "Not too many people want to drink heavy
water," Prins says.

Spalding began working with Arner, and by 2006 she had developed a regimen for isolating fat cells from the vast array of other
cells found in human flab. Analyzing fat biopsies and liposuction
leftovers from people of various ages, Spalding showed that people
born a few years before atomic bomb testing began had fat cells
with high levels of ^{14}C, which made sense only if these cells were
generated after the fallout had spiked the isotope's levels. When
Spalding looked at people born after the bomb tests, she saw fat
cells with different amounts of ^{14}C, levels corresponding to various
dates on the bomb-pulse curve. In all, the data indicate that people
replace half of their fat cells about every eight years, Spalding reported this summer in *Nature*.

"It's a landmark paper . . . . and a phenomenal advance on a number of fronts," says Prins. "You've got this technique out of *Star Trek*,
and now everybody thinks that fat is a dynamic organ." No drug
company would have looked into fat turnover before, he says, "but
now people will start to consider therapeutic perspectives."

Loving the Bomb

As the years go by, the ^{14}C level in the atmosphere is slowly returning to its prewar levels. Rising carbon dioxide emissions, chock-full
of ^{12}C, have only hastened the isotope's demise. And yet the bomb-
pulse technique is just taking off.

Both Frisén and Spalding, who left Frisén's lab in 2006 to become an assistant professor at Karolinska, are expanding its applications. Entering debates similar to the ones about neurogenesis
and fat turnover, they're looking at whether heart cells and insulin-
producing beta cells in the pancreas renew throughout life or
whether we're stuck with the ones we're born with. In tissues in
which stem cells have been identified, they plan to examine how often these cells divide and how they are made.

"The clinical implications are huge," says Yuval Dor, a cell biologist at the Hebrew University–Hadassah Medical School in Jerusalem, Israel, and an observer of the bomb-pulse technique. "There are hundreds of great biological questions that can be answered . . . We're all very much looking forward to how this will turn out."

The weight isn't all on Spalding and Frisén's shoulders. Other groups have begun to experiment with the technique as well. Like Frisén, diabetologists David Harlan and Shira Perl of the U.S. National Institute of Diabetes and Digestive and Kidney Diseases in Bethesda, Maryland, are using ^{14}C to measure turnover in beta cells. And Lawrence Livermore's Buchholz says he's been approached by a number of labs interested in everything from climate modeling (changing weather patterns are reflected in ^{14}C levels in coral) to dating confiscated ivory tusks and authenticating wine vintages.

Still, it's not a technique that most labs have the resources to adopt. "There are no kits you can buy to do this," says Buchholz. And most labs don't have access to the powerful isotope detectors needed to perform the ^{14}C analysis.

Critics also point out that the bomb-pulse technique has limitations. Although Spalding's work supported Rakic's stance on neurogenesis, Rakic notes that when damaged cells repair DNA, that DNA could incorporate new ^{14}C, suggesting new cell formation when there is none. Conversely, fat-turnover expert Prins says that new cells sometimes recycle DNA from dead cells, giving the impression — under ^{14}C analysis — that no new cells have been made.

And Spalding admits that the forensics applications have a shelf life: as ^{14}C levels recede to background in the atmosphere — Buchholz estimates a return to prebomb conditions by 2020 — it will become harder and harder to tell a corpse's year of death. But she's optimistic that as isotope detectors become more sensitive — she's working with Wild and Kutschera to help make this happen — police will be solving cases with the technique for years to come. Research on brain, fat, and other clinical topics won't be affected by the dissipation because scientists can turn to tissue samples banked over the decades after the bomb tests.

Back at Karolinska, Spalding, Frisén, and a few other collaborators have just formed a Center of Excellence to map the regenerative potential of the entire human body. Over the next ten years, they'll try to gauge the turnover of every cell type they can. "I love this technique," says Frisén. "We're having a lot of fun with it."

Next year, for a sabbatical, Spalding will head off to California, where she will look for new challenges while continuing her brain and fat research. Stay tuned for an upcoming paper on neurogenesis in the hippocampus — and some more surprises with fat turnover.

Meanwhile, at the birthplace of the atomic bomb in New Mexico, the retired Los Alamos National Laboratory scientist Donald Barr reflects on what Spalding and the other bomb-pulse researchers are doing. He's been at the lab for more than fifty years, keeping tabs on nuclear fallout in the atmosphere, and he still comes in a couple of days a week to chat isotopes with his former colleagues. The mushroom clouds from nuclear detonations do indeed have a silver lining, he says. "There are questions we can now answer because of that testing that scientists never thought about at the time."

STEPHEN S. HALL

Last of the Neanderthals

FROM *National Geographic*

IN MARCH OF 1994 some spelunkers exploring an extensive cave system in northern Spain poked their lights into a small side gallery and noticed two human mandibles jutting out of the sandy soil. The cave, called El Sidrón, lay in the midst of a remote upland forest of chestnut and oak trees in the province of Asturias, just south of the Bay of Biscay. Suspecting that the jawbones might date back as far as the Spanish Civil War, when Republican partisans used El Sidrón to hide from Franco's soldiers, the cavers immediately notified the local Guardia Civil. But when police investigators inspected the gallery, they discovered the remains of a much larger — and, it would turn out, much older — tragedy.

Within days, law enforcement officials had shoveled out some 140 bones, and a local judge ordered the remains sent to the national forensic pathology institute in Madrid. By the time scientists finished their analysis (it took the better part of six years), Spain had its earliest cold case. The bones from El Sidrón were not Republican soldiers but the fossilized remains of a group of Neanderthals who lived, and perhaps died violently, approximately 43,000 years ago. The locale places them at one of the most important geographical intersections of prehistory, and the date puts them squarely at the center of one of the most enduring mysteries in all of human evolution.

The Neanderthals, our closest prehistoric relatives, dominated Eurasia for the better part of 200,000 years. During that time, they poked their famously large and protruding noses into every corner of Europe and beyond — south along the Mediterranean from the Strait of Gibraltar to Greece and Iraq, north to Russia, as far west as

Britain, and almost to Mongolia in the east. Scientists estimate that even at the height of the Neanderthal occupation of western Europe, their total number probably never exceeded 15,000. Yet they managed to endure, even when a cooling climate turned much of their territory into something like northern Scandinavia today — a frigid, barren tundra, its bleak horizon broken by a few scraggly trees and just enough lichen to keep the reindeer happy.

By the time of the tragedy at El Sidrón, however, the Neanderthals were on the run, seemingly pinned down in Iberia, pockets of central Europe, and along the southern Mediterranean by a deteriorating climate, and further squeezed by the westward spread of anatomically modern humans as they emerged from Africa into the Middle East and beyond. Within another 15,000 years or so, the Neanderthals were gone forever, leaving behind a few bones and a lot of questions. Were they a clever and perseverant breed of survivors, much like us, or a cognitively challenged dead end? What happened during that period, roughly 45,000 to 30,000 years ago, when the Neanderthals shared some parts of the Eurasian landscape with those modern-human migrants from Africa? Why did one kind of human being survive and the other disappear?

On a damp, fog-shrouded morning in September 2007, I stood before the entrance to El Sidrón with Antonio Rosas of the National Museum of Natural Sciences in Madrid, who heads the paleoanthropological investigation. One of his colleagues handed me a flashlight, and I gingerly lowered myself into the black hole. As my eyes adjusted to the interior, I began to make out the fantastic contours of a karstic cave. An underground river had hollowed out a deep vein of sandstone, leaving behind a limestone cavern extending hundreds of yards, with side galleries spidering out to at least twelve entrances. Ten minutes into the cave, I arrived at the Galería del Osario — the "tunnel of bones." Since 2000 some 1,500 bone fragments have been unearthed from this side gallery, representing the remains of at least nine Neanderthals — five young adults, two adolescents, a child of about eight, and a three-year-old toddler. All showed signs of nutritional stress in their teeth — not unusual in young Neanderthals late in their time on Earth. But a deeper desperation is etched in their bones. Rosas picked up a recently unearthed fragment of a skull and another of a long bone of an arm, both with jagged edges.

"These fractures were — clop — made by humans," Rosas said,

imitating the blow of a stone tool. "It means these fellows went after the brains and into long bones for the marrow."

In addition to the fractures, cut marks left on the bones by stone tools clearly indicate that the individuals were cannibalized. Whoever ate their flesh, and for whatever reason — starvation? ritual? — the subsequent fate of their remains bestowed upon them a distinct and marvelous kind of immortality. Shortly after the nine individuals died — possibly within days — the ground below them suddenly collapsed, leaving little time for hyenas and other scavengers to scatter the remains. A slurry of bones, sediment, and rocks tumbled sixty feet into a hollow limestone chamber below, much the way mud fills the inside walls of a house during a flood.

There, buffered by sand and clay, preserved by the cave's constant temperature, and sequestered in their jewel cases of mineralized bone, a few precious molecules of the Neanderthals' genetic code survived, awaiting a time in the distant future when they could be plucked out, pieced together, and examined for clues to how these people lived and why they vanished.

The first clue that our kind of human was not the first to inhabit Europe turned up a century and a half ago about eight miles east of Düsseldorf, Germany. In August 1856 laborers quarrying limestone from a cave in the Neander Valley dug out a beetle browed skullcap and some thick limb bones. Right from the start, the Neanderthals were saddled with an enduring cultural stereotype as dim-witted, brutish cavemen. The size and shape of the fossils does suggest a short, stout fireplug of a physique (males averaged about five feet five inches tall and about 185 pounds), with massive muscles and a flaring rib cage presumably encasing capacious lungs. Steven E. Churchill, a paleoanthropologist at Duke University, has calculated that to support his body mass in a cold climate, a typical Neanderthal male would have needed up to 5,000 calories daily, or approaching what a bicyclist burns each day in the Tour de France. Yet behind its bulging browridge, a Neanderthal's low-domed skull housed a brain with a volume slightly larger on average than our own today. And while their tools and weapons were more primitive than those of the modern humans who supplanted them in Europe, they were no less sophisticated than the implements made by their modern-human contemporaries living in Africa and the Middle East.

One of the longest and most heated controversies in human evolution rages around the genetic relationship between Neanderthals and their European successors. Did the modern humans sweeping out of Africa beginning some 60,000 years ago completely replace the Neanderthals, or did they interbreed with them? In 1997 the latter hypothesis was dealt a powerful blow by geneticist Svante Pääbo — then at the University of Munich — who used an arm bone from the original Neanderthal man to deliver it. Pääbo and his colleagues were able to extract a tiny 378-letter snippet of mitochondrial DNA (a kind of short genetic appendix to the main text in each cell) from the 40,000-year-old specimen. When they read out the letters of the code, they found that the specimen's DNA differed from living humans to a degree suggesting that the Neanderthal and modern-human lineages had begun to diverge long before the modern-human migration out of Africa. Thus the two represent separate geographic and evolutionary branches splitting from a common ancestor. "North of the Mediterranean, this lineage became Neanderthals," said Chris Stringer, research leader on human origins at the Natural History Museum in London, "and south of the Mediterranean, it became us." If there was any interbreeding when they encountered each other later, it was too rare to leave a trace of Neanderthal mitochondrial DNA in the cells of living people.

Pääbo's genetic bombshell seemed to confirm that Neanderthals were a separate species — but it does nothing to solve the mystery of why they vanished and our species survived.

One obvious possibility is that modern humans were simply more clever, more sophisticated, more "human." Until recently, archaeologists pointed to a "great leap forward" around 40,000 years ago in Europe, when the Neanderthals' relatively humdrum stone tool industry — called Mousterian, after the Le Moustier site in southwestern France — gave way to the more varied stone and bone tool kits, body ornaments, and other signs of symbolic expression associated with the appearance of modern humans. Some scientists, such as the Stanford University anthropologist Richard Klein, still argue for some dramatic genetic change in the brain — possibly associated with a development in language — that propelled early modern humans to cultural dominance at the expense of their beetle-browed forebears.

But the evidence in the ground is not so cut and dried. In 1979 archaeologists discovered a late Neanderthal skeleton at Saint-Césaire in southwestern France surrounded not with typical Mousterian implements but with a surprisingly modern repertoire of tools. In 1996 Jean-Jacques Hublin of the Max Planck Institute in Leipzig and Fred Spoor of University College London identified a Neanderthal bone in another French cave, near Arcy-sur-Cure, in a layer of sediment also containing ornamental objects previously associated only with modern humans, such as pierced animal teeth and ivory rings. Some scientists, such as the British paleoanthropologist Paul Mellars, dismiss such modern "accessorizing" of a fundamentally archaic lifestyle as an "improbable coincidence" — a last gasp of imitative behavior by Neanderthals before the inventive newcomers out of Africa replaced them. But more recently, Francesco d'Errico of the University of Bordeaux and Marie Soressi, also at the Max Planck Institute in Leipzig, analyzed hundreds of crayonlike blocks of manganese dioxide from a French cave called Pech de l'Azé, where Neanderthals lived well before modern humans arrived in Europe. D'Errico and Soressi argue that the Neanderthals used the black pigment for body decoration, demonstrating that they were fully capable of achieving "behavioral modernity" all on their own.

"At the time of the biological transition," says Erik Trinkaus, a paleoanthropologist at Washington University in St. Louis, "the basic behavior [of the two groups] is pretty much the same, and any differences are likely to have been subtle." Trinkaus believes they may indeed have mated occasionally. He sees evidence of admixture between Neanderthals and modern humans in certain fossils, such as a 24,500-year-old skeleton of a young child discovered at the Portuguese site of Lagar Velho and a 32,000-year-old skull from a cave called Muierii in Romania. "There were very few people on the landscape, and you need to find a mate and reproduce," says Trinkaus. "Why not? Humans are not known to be choosy. Sex happens."

It may have happened, other researchers say, but not often, and not in a way that left behind any evidence. Katerina Harvati, another researcher at the Max Planck Institute in Leipzig, has used detailed 3-D measurements of Neanderthal and early modern-human fossils to predict exactly what hybrids between the two would have looked like. None of the fossils examined so far matches her predictions.

The disagreement between Trinkaus and Harvati is hardly the first time that two respected paleoanthropologists have looked at the same set of bones and come up with mutually contradictory interpretations. Pondering — and debating — the meaning of fossil anatomy will always play a role in understanding Neanderthals. But now there are other ways to bring them back to life.

Two days after my first descent into El Sidrón cave, Araceli Soto Flórez, a graduate student at the University of Oviedo, came across a fresh Neanderthal bone, probably a fragment of a femur. All digging immediately ceased, and most of the crew evacuated the chamber. Soto Flórez then squeezed herself into a sterile jumpsuit, gloves, booties, and plastic face mask. Under the watchful eyes of Antonio Rosas and the molecular biologist Carles Lalueza-Fox, she delicately extracted the bone from the soil, placed it in a sterile plastic bag, and deposited the bag in a chest of ice. After a brief stop in a hotel freezer in nearby Villamayo, the leg bone eventually arrived at Lalueza-Fox's laboratory at the Institute of Evolutionary Biology in Barcelona. His interest was not in the anatomy of the leg or anything it might reveal about Neanderthal locomotion. All he wanted from it was its DNA.

Prehistoric cannibalism has been very good for modern-day molecular biology. Scraping flesh from a bone also removes the DNA of microorganisms that might otherwise contaminate the sample. The bones of El Sidrón have not yielded the most DNA of any Neanderthal fossil — that honor belongs to a specimen from Croatia, also cannibalized — but so far they have revealed the most compelling insights into Neanderthal appearance and behavior. In October 2007 Lalueza-Fox, Holger Römpler of the University of Leipzig, and their colleagues announced that they had isolated a pigmentation gene from the DNA of an individual at El Sidrón (as well as another Neanderthal fossil from Italy). The particular form of the gene, called *MC1R*, indicated that at least some Neanderthals would have had red hair, pale skin, and, possibly, freckles. The gene is unlike that of red-haired people today, however — suggesting that Neanderthals and modern humans developed the trait independently, perhaps under similar pressures in northern latitudes to evolve fair skin to let in more sunlight for the manufacture of vitamin D. Just a few weeks earlier, Svante Pääbo, who now heads the genetics laboratory at the Max Planck Institute in Leipzig, Lalueza-

Fox, and their colleagues had announced an even more astonishing find: two El Sidrón individuals appeared to share with modern humans a version of a gene called *FOXP2* that contributes to speech and language ability, acting not only in the brain but also on the nerves that control facial muscles. Whether Neanderthals were capable of sophisticated language abilities or a more primitive form of vocal communication (singing, for example) still remains unclear, but the new genetic findings suggest they possessed some of the same vocalizing hardware as modern humans.

All this from a group of ill-fated Neanderthals buried in a cave collapse soon after they were consumed by their own kind.

"So maybe it's a good thing to eat your conspecifics," says Pääbo.

A tall, cheerful Swede, Pääbo is the main engine behind a breathtaking scientific tour de force: the attempt, expected to be completed next month, to read out not just single Neanderthal genes but the entire 3-billion-letter sequence of the Neanderthal genome. Traces of DNA in fossils are vanishingly faint, and because Neanderthal DNA is ever so close to that of living people, one of the biggest hurdles in sequencing it is the ever-present threat of contamination by modern-human DNA — especially by the scientists handling the specimens. The precautions taken in excavating at El Sidrón are now becoming standard practice at other Neanderthal sites. Most of the DNA for Pääbo's genome project, however, has come from the Croatian specimen, a 38,000-year-old fragment of leg bone found almost thirty years ago in the Vindija cave. Originally deemed unimportant, it sat in a drawer in Zagreb, largely untouched and thus uncontaminated, for most of its museum life.

Now it is the equivalent of a gold mine for prehistoric human DNA, albeit an extremely difficult mine to work. After the DNA is extracted in a sterile laboratory in the basement of the Max Planck Institute, it is shipped overnight to Branford, Connecticut, where collaborators at 454 Life Sciences have invented machines that can rapidly decipher the sequence of DNA's chemical letters. The vast majority of those letters spell out bacterial contaminants or other non-Neanderthal genetic information. But in the fall of 2006, Pääbo and his colleagues announced they had deciphered approximately one million letters of Neanderthal DNA. (At the same time, a second group, headed by Edward Rubin at the Department of Energy Joint Genome Institute in Walnut Creek, California, used DNA provided by Pääbo to read out snippets of genetic code using a different approach.) By last year, dogged by claims that their work had serious contamination problems, the Leipzig group claimed to have improved its accuracy and identified about 70 million letters of DNA — roughly 2 percent of the total.

"We know that the human and chimpanzee sequences are 98.7 percent the same, and Neanderthals are much closer to us than chimps," said Ed Green, head of biomathematics in Pääbo's group in Leipzig, "so the reality is that for most of the sequence, there's no difference between Neanderthals and [modern] humans." But the differences — less than 0.5 percent of the sequence — are enough to confirm that the two lineages had begun to diverge around 700,000 years ago. The Leipzig group also managed to extract mitochondrial DNA from two fossils of uncertain origin that had been excavated in Uzbekistan and southern Siberia; both had a uniquely Neanderthal genetic signature. While the Uzbekistan specimen, a young boy, had long been considered a Neanderthal, the Siberian specimen was a huge surprise, extending the known Neanderthal range some 1,200 miles east of their European stronghold.

So, while the new genetic evidence appears to confirm that Neanderthals were a separate species from us, it also suggests that they may have possessed human language and were successful over a far larger sweep of Eurasia than previously thought. Which brings us back to the same hauntingly persistent question that has shadowed them from the beginning: why did they disappear?

To coax a Neanderthal fossil to reveal its secrets, you can measure it with calipers, probe it with a CT scan, or try to capture the ghost of its genetic code. Or if you happen to have at your disposal a type of particle accelerator called a synchrotron, you can put it in a lead-lined room and blast it with a 50,000-volt X-ray beam without disturbing so much as a single molecule.

Over a sleep-deprived week in October 2007, a team of scientists gathered at the European Synchrotron Radiation Facility (ESRF) in Grenoble, France, for an unprecedented "convention of jawbones." The goal was to explore a crucial question in the life history of the Neanderthals: did they reach maturity at an earlier age than their modern-human counterparts? If so, it might have implications for their brain development, which in turn might help explain why they disappeared. The place to look for answers was deep inside the structure of Neanderthal teeth.

"When I was young, I thought that teeth were not so useful in assessing recent human evolution, but now I think they are the most important thing," said Jean-Jacques Hublin, who had accompanied his Max Planck Institute colleague Tanya Smith to Grenoble.

Along with Paul Tafforeau of the ESRF, Hublin and Smith were squeezed into a computer-filled hutch at the facility — one of the three largest synchrotrons in the world, with a storage ring for energized electrons half a mile in circumference — watching on a video monitor as the X-ray beam zipped through the right upper canine of an adolescent Neanderthal from the site of Le Moustier in southwestern France, creating arguably the most detailed dental X-ray in human history. Meanwhile, a dream team of other fossils sat on a shelf nearby, awaiting their turn in the synchrotron's spotlight: two jawbones of Neanderthal juveniles recovered in Krapina, Croatia, dating back 130,000 to 120,000 years; the so-called La Quina skull from a Neanderthal youth, discovered in France and dating from between 75,000 to 40,000 years ago; and two striking 90,000-year-old modern-human specimens, teeth intact, found in a rock shelter called Qafzeh in Israel.

When teeth are imaged at high resolution, they reveal a complex, three-dimensional hatch of daily and longer periodic growth lines, like tree rings, along with stress lines that encode key moments in an individual's life history. The trauma of birth etches a sharp neonatal stress line on the enamel; the time of weaning and episodes of nutritional deprivation or other environmental stresses similarly leave distinct marks on developing teeth. "Teeth preserve a continuous, permanent record of growth, from before birth until they finish growing at the end of adolescence," Smith explained. Human beings take longer to reach puberty than chimpanzees, our nearest living relatives — which means more time spent learning and developing within the context of the social group. Early hominin species that lived on the savanna in Africa millions of years ago matured fast, more like chimps. So when in evolution did the longer modern pattern begin?

To address this question, Smith, Tafforeau, and colleagues had previously used the synchrotron to demonstrate that an early modern-human child from a site called Jebel Irhoud in Morocco (dated to around 160,000 years ago) showed the modern-human life history pattern. In contrast, the "growth rings" in the 100,000-year-old tooth of a young Neanderthal discovered in the Scladina cave in Belgium indicated that the child was eight years old when it died and appeared to be on track to reach puberty several years sooner than the average for modern humans. Another research team, using a single Neanderthal tooth, had found no such difference between its growth pattern and that of living humans. But while a full analysis from the "jawbone convention" would take time, preliminary results, Smith said, were "consistent with what we see in Scladina."

"This would certainly affect Neanderthal social organization, mating strategy, and parenting behavior," says Hublin. "Imagine a society where individuals start to reproduce four years earlier than in modern humans. It's a very different society. It could also mean the Neanderthals' cognitive abilities may have been different from modern humans'."

Neanderthal society may have differed in another way crucial to group survival, in what archaeologists call cultural buffering. A buffer is something in a group's behavior — a technology, a form of social organization, a cultural tradition — that hedges its bets in the high-stakes game of natural selection. It's like having a small cache of extra chips at your elbow in a poker game, so you don't have to fold your hand quite as soon. For example, Mary Stiner and Steven Kuhn of the University of Arizona argue that early modern humans emerged from Africa with the buffer of an economically efficient approach to hunting and gathering that resulted in a more diverse diet. While men chased after large animals, women and children foraged for small game and plant foods. Stiner and Kuhn maintain that Neanderthals did not enjoy the benefits of such a marked division of labor. From southern Israel to northern Germany, the archaeological record shows that Neanderthals instead relied almost entirely on hunting big and medium-size mammals like horses, deer, bison, and wild cattle. No doubt they were eating some vegetable material and even shellfish near the Mediterranean, but the lack of milling stones or other evidence for processing plant foods suggests to Stiner and Kuhn that to a Neanderthal vegetables were supplementary foods, "more like salads, snacks, and desserts than energy-rich staple foods."

Their bodies' relentless demand for calories, especially in higher

latitudes and during colder interludes, probably forced Neanderthal women and children to join in the hunt — a "rough and dangerous business," write Stiner and Kuhn, judging by the many
healed fractures evident on Neanderthal upper limbs and skulls.
The modern-human bands that arrived on the landscape toward
the end of the Neanderthals' time had other options.

"By diversifying diet and having personnel who [did different
tasks], you have a formula for spreading risk, and that is ultimately
good news for pregnant women and for kids," Stiner told me. "So if
one thing falls through, there's something else." A Neanderthal
woman would have been powerful and resilient. But without such
cultural buffering, she and her young would have been at a disadvantage.

Of all possible cultural buffers, perhaps the most important was
the cushion of society itself. According to Erik Trinkaus, a Neanderthal social unit would have been about the size of an extended
family. But in early modern-human sites in Europe, Trinkaus said,
"we start getting sites that represent larger populations." Simply living in a larger group has biological as well as social repercussions.
Larger groups inevitably demand more social interactions, which
goads the brain into greater activity during childhood and adolescence, creates pressure to increase the sophistication of language,
and indirectly increases the average life span of group members.
Longevity, in turn, increases intergenerational transmission of
knowledge and creates what Chris Stringer calls a "culture of inno-
vation" — the passage of practical survival skills and toolmaking
technology from one generation to the next, and later between
one group and another.

Whatever the suite of cultural buffers, they may well have provided an extra, albeit thin, layer of insulation against the harsh cli-
matic stresses that Stringer argues peaked right around the time
the Neanderthals vanished. Ice-core data suggest that from about
30,000 years ago until the last glacial maximum about 18,000 years
ago, Earth's climate fluctuated wildly, sometimes within the space
of decades. A few more people in the social unit, with a few more
skills, might have given modern humans an edge when conditions
turned harsh. "Not a vast edge," Stringer said. "Neanderthals were
obviously well adapted to a colder climate. But with the superimpo-
sition of these extreme changes in climate on the competition with
modern humans, I think that made the difference."

Which leaves the final, delicate — and, as Jean-Jacques Hublin likes to say, politically incorrect — question that has bedeviled Neanderthal studies since the Out-of-Africa theory became generally accepted: was the replacement by modern humans attenuated and peaceful, the Pleistocene version of kissing cousins, or was it relatively swift and hostile?

"Most Neanderthals and modern humans probably lived most of their lives without seeing each other," he said, carefully choosing his words. "The way I imagine it is that occasionally in these border areas, some of these guys would see each other at a distance . . . but I think the most likely thing is that they excluded each other from the landscape. Not just avoided, but excluded. We know from recent research on hunter-gatherers that they are much less peaceful than generally believed."

"Sometimes I just turn out the lights in here and think what it must have been like for them."

The evolutionary biologist Clive Finlayson, of the Gibraltar Museum, was standing in the vestibule of Gorham's Cave, a magnificent tabernacle of limestone opening to the sea on the Rock of Gibraltar. Inside, fantastic excretions of flowstone drooled from the ceiling of the massive nave. The stratigraphy in the cave is pocked with evidence of Neanderthal occupation going back 125,000 years, including stone spear points and scrapers, charred pine nuts, and the remains of ancient hearths. Two years ago, Finlayson and his colleagues used radiocarbon dating to determine that the embers in some of those fireplaces died out only 28,000 years ago — the last known trace of Neanderthals on Earth. (Other hearths in the cave may be as young as 24,000 years old, but their dating is controversial.)

From pollen and animal remains, Finlayson has reconstructed what the environment was like from 50,000 to 30,000 years ago. Back then a narrow coastal shelf surrounded Gibraltar, the Mediterranean two or three miles distant. The landscape was scrub savanna scented with rosemary and thyme, its rolling sand dunes interrupted by the occasional cork oak and stone pine, with wild asparagus growing in the coastal flats. Prehistoric vultures, some with nine-foot wingspans, nested high up in the cliff face, scanning the dunes for meals. Finlayson imagines the Neanderthals watching the birds circle and descend, then racing them for food. Their diet was certainly more varied than the typical Neanderthal dependence on terrestrial game. His research team has found rabbit bones, tortoise shells, and mussels in the cave, along with dolphin bones and a seal skeleton with cut marks. "Except for rice, you've almost got a Mousterian paella!" Finlayson joked.

But then things changed. When the coldest fingers of the ice age finally reached southern Iberia in a series of abrupt fluctuations between 30,000 and 23,000 years ago, the landscape was transformed into a semiarid steppe. On this more open playing field, perhaps the tall, gracile modern humans moving into the region with projectile spears gained the advantage over the stumpy, musclebound Neanderthals. But Finlayson argues that it was not so much the arrival of modern humans as the dramatic shifts in climate that pushed the Iberian Neanderthals to the brink. "A three-year period of intense cold, or a landslide, when you're down to ten people, could be enough," he said. "Once you reach a certain level, you're the living dead."

The larger point may be that the demise of the Neanderthals is not a sprawling yet coherent paleoanthropological novel; rather, it is a collection of related but unique short stories of extinction. "Why did the Neanderthals disappear in Mongolia?" Stringer asked. "Why did they disappear in Israel? Why did they disappear in Italy, in Gibraltar, in Britain? Well, the answer could be different in different places, because it probably happened at different times. So we're talking about a large range, and a disappearance and retreat at different times, with pockets of Neanderthals no doubt surviving in different places at different times. Gibraltar is certainly one of their last outposts. It could be the last, but we don't know for sure."

Whatever happened, the denouement of all these stories had a signatory in Gorham's Cave. In a deep recess of the cavern, not far from that last Neanderthal hearth, Finlayson's team recently discovered several red handprints on the wall, a sign that modern humans had arrived in Gibraltar. Preliminary analysis of the pigments dates the handprints between 20,300 and 19,500 years ago. "It's like they were saying, Hey, it's a new world now," said Finlayson.

SUE HALPERN

Virtual Iraq

FROM *The New Yorker*

IN NOVEMBER 2004, when he was nineteen years old, a marine I'll call Travis Boyd found himself about to rush the roof of the tallest building in the northern end of Falluja in the midst of a firefight. Boyd, whose first assignment in Iraq was to the security detail at Abu Ghraib prison, had been patrolling the city with his thirteen-man infantry squad, rooting out insurgents and sleeping on the floors of abandoned houses, where they'd often have to remove dead bodies in order to lay out their bedrolls.

With Boyd in the lead, the marines ran up the building's four flights of stairs. When they reached the top, "the enemy cut loose at us with everything they had," he recalled. "Bullets were exploding like firecrackers all around us." Boyd paused, and his team leader, whom he thought of as an older brother, ran past him to the far side of the building. Moments after he got there, the team leader was shot dead. Within minutes, everyone else on the roof was wounded. "We had to crawl out of there," said Boyd, who was hit with shrapnel and suffered a concussion, earning a Purple Heart. "That was my worst day."

It is in the nature of soldiers to put emotions aside, and that is what Boyd did for three years. He "stayed on the line" with his squad and finished his tour of duty the following June, married his high school girlfriend, and soon afterward began training for his second Iraq deployment, not thinking much about what he had seen or done during the first. Haditha, where he was sent in the fall of 2005, was calmer than Falluja. There were roadside bombs but no direct attacks. Boyd was now a team leader, and he and his men patrolled the streets like police. When drivers did not respond to the soldiers' efforts to get them to stop, he said, "we'd have to light them up." He was there for seven months.

With one more year of service left on his commitment and not enough time for a third deployment, Boyd was separated from his unit and assigned to fold towels and clean equipment at the fitness center of his stateside base. It was a quiet, undemanding job, intended to allow him to decompress from combat. Instead, he was haunted by memories of Iraq. He couldn't sleep. His mind raced. He was edgy, guilt-racked, depressed. He could barely do his job.

"I'd avoid crowds, I'd avoid driving, I'd avoid going out at night," he told me the first time we spoke. "I'd avoid people who weren't infantry, the ones who hadn't been bleeding and dying and going weeks and months without showers and eating MREs. I'd have my wife drive me if I had to go off the base. A few times I thought I saw a mortar in the road and reached for the steering wheel. I was always on alert, ready for anything to happen at any time."

Eventually, as part of a standard medical screening, Boyd was diagnosed as having chronic posttraumatic stress disorder. PTSD, which in earlier conflicts was known as battle fatigue or shell shock but is not exclusively war-related, has been an officially recognized medical condition since 1980, when it entered the American Psychiatric Association's *Diagnostic and Statistical Manual of Mental Disorders*. (In an earlier edition, it was called "gross stress reaction.") PTSD is precipitated by a terrifying event or situation — war, a car accident, rape, planes crashing into the World Trade Center — and is characterized by nightmares, flashbacks, and intrusive and uncontrollable thoughts, as well as by emotional detachment, numbness, jumpiness, anger, and avoidance. Boyd's doctor prescribed medicine for his insomnia and encouraged him to seek out psychotherapy, telling him about an experimental treatment option called Virtual Iraq, in which patients worked through their combat trauma in a computer-simulated environment. The portal was a head-mounted display (a helmet with a pair of video goggles), earphones, a scent-producing machine, and a modified version of Full Spectrum Warrior, a popular video game.

When Travis Boyd agreed to become a subject in the Virtual Iraq clinical trial, in the spring of 2007, he became one of about thirty-five active-duty and former members of the military to use the program to treat their psychological wounds. Currently, the Department of Defense is testing Virtual Iraq — one of three virtual-reality programs it has funded for PTSD treatment, and the only one aimed at "ground pounders" like Boyd — in six locations, including the Naval Medical Center San Diego, Walter Reed Army Medical Center, in Washington, D.C., and Weill Cornell Medical College, in New York. According to a recent study by the RAND Corporation, nearly 20 percent of Iraq and Afghanistan war veterans are suffering from PTSD or major depression. Almost half won't seek treatment. If virtual-reality exposure therapy proves to be clinically validated — only preliminary results are available so far — it may be more than another tool in the therapists' kit: it may encourage those in need to seek help.

"Most PTSD therapies that we've seen don't seem to be working, so what's the harm in dedicating some money to R & D that might prove valuable?" Paul Rieckhoff, the executive director of Iraq and Afghanistan Veterans of America, said last November. In January his group issued a lengthy report called "Mental Health Injuries: The Invisible Wounds of War," which cited research suggesting that "multiple tours and inadequate time at home between deployments increase rates of combat stress by 50%." Rieckhoff went on, "I'm not someone who responds to sitting with some guy, talking about my whole life. I'm going to go in and talk to some dude who doesn't understand my shit and talk about my mom? I'm the worst of that kind of guy. So VR therapy, maybe it will work. We're a video-game generation. It's what we grew up on. So maybe we'll respond to it."

Strictly speaking, using virtual reality to treat combat-related PTSD is not new. In 1997, more than twenty years after the Vietnam War ended, researchers in Atlanta unveiled Virtual Vietnam. It dropped viewers into one of two scenarios: a jungle clearing with a "hot" landing zone or a Huey helicopter, its rotors whirring, its body casting a running shadow over rice paddies, a dense tropical forest, and a river. The graphics were fairly crude, and the therapist had a limited number of sights and sounds to manipulate, but Virtual Vietnam had the effect of putting old soldiers back in the thick of war. Ten combat veterans with long-term PTSD who had not responded to multiple interventions participated in a clinical trial of Virtual Vietnam, typically lasting a month or two. All of them showed significant signs of improvement, both directly after treatment and in a follow-up half a year later. (PTSD is assessed on a number of scales, some subjective and others based on the observation of the clinician.) As successful as it was, though, Virtual Vietnam didn't catch on. It was an experiment, and when the experiment was over the researchers moved on.

Like Virtual Vietnam, Virtual Iraq is a tool for doing what's known as prolonged-exposure therapy, which is sometimes called immersion therapy. It is a kind of cognitive-behavioral therapy (CBT), derived from Pavlov's classic work with dogs. Prolonged-exposure therapy, which falls under the rubric of CBT, is at once intuitively obvious and counterintuitive: it requires the patient to revisit and retell the story of the trauma over and over again and, through a psychological process called "habituation," rid it of its overwhelming power. The idea is to disconnect the memory from the reactions to the memory, so that although the memory of the traumatic event remains, the everyday things that can trigger fear and panic, such as trash blowing across the interstate or a car backfiring — what psychologists refer to as cues — are restored to insignificance. The trauma thus becomes a discrete event, not a constant, self-replicating, encompassing condition.

This process was explained to me by JoAnn Difede, the director of the Program for Anxiety and Traumatic Stress Studies at Weill Cornell, when I visited her in her office last fall. Difede, a tough-minded New Yorker, began using virtual-reality exposure therapy with patients from the hospital's burn unit in the 1990s. She treated victims of September 11 with a program called Virtual WTC, which she designed with the creators of Virtual Vietnam, and is currently running a Virtual Iraq clinical trial as well as supervising therapists at other study sites.

Difede says that therapists have been slow to adopt exposure therapy because they worry that it might be cruel to immerse a patient in a drowning pool of painful memories. It's a worry that, she believes, misses the point of the therapy. "If you suddenly become afraid of the staircase because you had to walk down twenty-five flights of stairs to get out of the World Trade Center, the stairs went from being neutral to being negative," Difede explained. "What we should be doing is extinguishing the cues associated with the stimuli, which should allow for a more complete remission, as well as mastery of the experience. It also should allow for greater emotional engagement. Because numbing and avoidance are symptoms of PTSD, you're asking the person to do in treatment the very thing their mind is avoiding doing. That's quite a dilemma."

It's this dilemma that makes virtual reality especially attractive to clinical psychologists like Difede. Because the traumatic environment is produced in a computer graphics lab, and its elements are controlled by the therapist, virtual reality can nudge an imagination that is at once overactive and repressed. "Voilà, you're there!" Difede said. "You don't have to do any work. You don't have to engage in any mental effort. We'll do it for you. We'll bring you there and then, gradually, we'll let you get involved in the experience in sensory detail."

When Travis Boyd was first asked to consider enrolling in the Virtual Iraq clinical trial, he was hesitant. He had already decided not to talk to his division therapist because "I didn't want to have it on my military record that I was crazy," he said. And he was a marine. "Infantry is supposed to be the toughest of the tough. Even though there was no punishment for going to therapy, it was looked down upon and seen as weak. But VR sounded pretty cool. They hook you up to a machine and you play around like a video game." Telling his buddies that he was going off to do VR was a lot easier than telling them he was seeing a shrink.

Before he was introduced to Virtual Iraq, the therapist asked him to close his eyes and talk about his wartime experiences. Without much prompting, he was back on the roof in Falluja, under fire, stalled at the top of the stairs, watching his friend and team leader run past him and die, and then he was dragging out his friend's body, looking at his messed-up face. When Boyd was finished, the doctor asked him to tell the story again. And, when he was finished that time, to tell it again. As he did, she asked him what he was smelling, and if the enemy was on the roof opposite or on the roof next door, and if there were planes overhead. She wanted to learn the details of his narrative and determine which moments were most troubling to him — she called them "hot spots" — and to figure out how she was going to use the sensory variables embedded in Virtual Iraq.

Boyd was introduced to the VR program in the third session. (There were twelve sessions in all, each about two hours long, over a period of six weeks.) Virtual-reality exposure therapy immerses the patient gradually; that first time Boyd just sat there with the VR gear on, looking at an Iraqi street scene, getting acquainted with the virtual world. Sound, which psychologists believe may stimulate memory more effectively than sight does, was added next and, with it, touch. "I'm talking about the firefight and she turns on this vibrating thing so you feel like you're in a shaking building," Boyd said. "Each time she added something, like an IED going off or a plane flying over, I'd become more emotional. We'd do it over and over, and it would become easier, and then she'd add something more, and the same thing would happen. I'd talk for forty minutes about this one five-minute thing. When it's only visual, it's not really real — it's just a video game — but when the ground starts vibrating and you smell smoke and hear the AK-47 firing, it becomes very real. I'd be shaking. When it was over, I'd go home and cry."

The inventor of Virtual Iraq is Albert Rizzo, a clinical psychologist at the University of Southern California, who goes by the nickname Skip. Rizzo, who is fifty-three, has thinning black hair that's down to his shoulders when it's not pulled back in a ponytail, a stud earring, and a nose that looks like it has met a boot or two — he plays rugby. Rizzo rides a Harley 1200 Sportster ("It's not a girl's bike, no matter what anyone tells you"), plays blues harmonica (he taught himself a couple of years ago, in order to reduce stress when he was commuting daily in LA traffic), and has an affable, jeans-and-untucked-shirt way about him that is particularly noticeable when he walks through Walter Reed or the Naval Medical Center San Diego alongside his starched military counterparts. In 2003, not long after the United States invaded Iraq, Rizzo, who was affiliated with the Institute for Creative Technologies, a USC offshoot that he likes to call "an unholy alliance between academia, Hollywood, and the military," had been designing virtual-reality systems to diagnose attention deficits in children and memory problems in older adults. But he had a hunch that if the war went on for very long, its veterans were going to come home with serious emotional problems.

"I thought we should be on this so we don't have another Vietnam, with all these guys suffering from PTSD," he told me one day last fall at Walter Reed, before he was to give a presentation to senior military officers. "I was working on a talk about virtual reality, just sniffing around the Internet, and I saw this link for the video game Full Spectrum Warrior." The game had, in fact, originated as a training device that the institute had developed for the Department of Defense. "I said, 'Oh, my God, that's Iraq!' It was instant. I thought we should take this game and run it in a head-mounted display right out of the box, for therapy."

Rizzo got in touch with Jarrell Pair, who had been the programmer on Virtual Vietnam, and convinced him to sign on to his as yet unfunded venture. By February 2004, he and Pair had built a prototype of Virtual Iraq on a laptop, using a single street in an Iraqi market town that they had recycled from Full Spectrum Warrior. To this they added a few alternate realities that a therapist could insert with a keystroke — a change from day to night, for example, or a switch from a deserted street to one where burka-clad shoppers strolled down the sidewalk. "That was our demo," Rizzo said. "We applied for money and we got nuked. Then the Hoge article comes out and everything changes overnight."

The article to which Rizzo was referring was written by Charles Hoge and his colleagues in the Department of Psychiatry and Behavioral Sciences at Walter Reed and was published in the *New England Journal of Medicine* that summer. It was the first assessment of mental-health problems emerging from service in Iraq and Afghanistan, and even its conservative estimate — that around 16 or 17 percent of those who fought in Iraq and 11 percent who served in Afghanistan were suffering from PTSD symptoms (an estimate that four years later has been revised dramatically upward) — caught the public and the military by surprise. Then Rizzo got a call from somebody in the Office of Naval Research. "He says, 'I hear you've got a prototype of Full Spectrum Warrior for PTSD,'" Rizzo recalled. "'We're going to try to get it funded.'"

The money came through in March 2005, and by the next fall, right around the time that Travis Boyd was being deployed to police Haditha, the first patients were recruited to try it out.

Before Skip Rizzo started designing virtual-reality systems, he was a conventional clinical psychologist, schooled in a variety of therapeutic methods. Rizzo grew up just outside Hartford, attended the

University of Hartford as an undergraduate, received a doctorate from Binghamton University, and did his internship at the VA hospital in Long Beach, California, not far from where he now lives. Then he took a job as a cognitive-rehabilitation therapist at a hospital in Costa Mesa, working with people who had suffered traumatic brain injuries. "A lot of young males are in that population," he said. "The high-risk-takers. The drunk drivers. Gang members — all of that. With that population, it was sometimes hard to motivate them to do the standard paper-and-pencil drill and practice routines. Then, in the early 1990s, Game Boys came on the scene, and it seemed to me that all my male clients, at every break, at every meal, had become Tetris warlords. It showed me that they were motivated to do game tasks and that the more they did them the better they got, and it hit me that there could be a link between cognitive rehabilitation and virtual reality." Rizzo left his job and accepted a postdoc at the Alzheimer Disease Research Center at USC, where he began to design rudimentary virtual-reality systems with the help of programmers in the computer science department. At the end of the postdoc, he moved to the engineering school at USC and started "building this stuff like crazy."

To make Virtual Iraq, Rizzo started with two basic scenarios: the market-town street scene and a Humvee moving along an Iraqi highway where all the exit signs are in Arabic and the road cuts through sand dunes. Then he gave therapists a menu of ways — visual, aural, tactile, even olfactory — to customize them. At the click of a mouse, the therapist can put the patient in the driver's seat of the Humvee, in the passenger's seat, or in the turret behind a machine gun, and the vehicle moves at a speed determined by the patient. Maybe the gunner in the turret is wearing night-vision goggles — the landscape goes grainy and green. A sandstorm could be raging (the driver can turn on the windshield wipers and beat it back); a dog could be barking; the inside of the vehicle could be rank. Rizzo's idea is that giving the therapist so many options — dusk, midday; with snipers, without snipers; driving fast, creeping along; the sound of a single mortar, the sound of multiple mortars; the sound of people yelling in English or in Arabic — increases the likelihood of evoking the patient's actual experience while engaging the patient on so many sensory levels that the immersion in the environment is nearly absolute.

"Tell me what you want me to add, anything," I overheard Rizzo asking a therapist at Walter Reed in February, a few days after she had completed a fourteen-session Virtual Iraq protocol in three months with the first soldier at the facility enrolled in the trial. (The patient didn't think he had gotten much better, though he was able to ride the subway again and no longer avoided large crowds.) "You're the one in the trenches hearing the stories. We'll keep evolving this to make it more relevant. What do you think about adding the smell of burning hair?"

Rizzo was sitting in a tiny, windowless room in front of a table ringed by a cloth skirt that partly hid the electrodes and other equipment that monitor a person's blood pressure, respiration, heart rate, and stress level during treatment and were connected to two computers. He had flown in the night before to install the latest software upgrade, which he was introducing to the therapist, a slight young woman in her thirties.

"OK," Rizzo said as he clicked the computer mouse rapidly, "this is really cool." On the screen was the basic Virtual Iraq market scene: a few nearly empty vendor stalls in the middle of a plaza and a row of small ground-floor shops in dun-colored buildings lining the sidewalk. "You walk to the end of this street" — the sound of footsteps could be heard — "it's market east. Now, let's see if this works. Let me blow up this car." He clicked again and a small car about the size of a Toyota Corolla, which had been parked at the curb, burst into flames. "It's a good effect. Now, when you blow up the car, put in 'add stunned civilian.' One more thing — you have to learn where the RPG guys are." He was referring to figures toting rocket-propelled grenades. "There's one here," he said, and on the screen there was another explosion. "Now we're going to head over there," he said, moving forward — more footsteps — toward a set of stairs. "Here's the deal with going up the stairs. You've got to hit it square on, otherwise you'll get caught up in the collision barrier. It just breaks the presence. You'll have to guide them. From here, there's a variety of things you can do. First off, you've got the insurgent on the roof over there. The insurgents just pop up. You have to learn where they are, too."

The therapist looked over Rizzo's shoulder while he brought a Black Hawk helicopter in for a flyover and then blew up another car on the street. "One thing I have to be careful about is not hitting something by accident," she said. "One time I mistakenly clicked my mouse and all of a sudden a bullet came flying out, and I had to tell the patient that I was sorry and didn't mean to do that."

The first time I put on a head-mounted display and headphones and entered Virtual Iraq had been in this same room at Walter Reed a few months earlier, after Rizzo presented preliminary results from a study site to a small gathering of military officials. Rizzo was having trouble linking his laptop's PowerPoint presentation to the Walter Reed audiovisual system, and he had to speak without notes, often from a crouch behind the podium as he picked through a jumble of cables searching for one that was live. "The last one hundred years, we've studied psychology in the real world," Rizzo told the group. "In the next hundred, we're going to study it in the virtual world." He threw out some numbers. Of the five subjects who had completed treatment, four no longer met the diagnostic criteria for PTSD. A fifth soldier showed no gain. (To these he would add, a few months later, the results for ten others, eight of whom had gotten better. Of the six research sites, San Diego was the first to have preliminary results.) After talking more generally about the features of Virtual Iraq, Rizzo invited everyone present to the fourth-floor psychiatric wing to try it out.

Although I had seen Virtual Iraq in one dimension on a computer monitor, encountering it in three dimensions, with my eyes blinkered by the headset and my ears getting a direct audio feed, was different. It still felt like make-believe, but I was fully engaged. Rizzo placed a dummy M4 rifle in my hands and guided my fingers to a video controller fixed to the barrel. (By design, patients who use Virtual Iraq do not fire weapons; the M4 is a mood-setting device, for verisimilitude.) One toggle moved me forward, another moved me back, and a third sped me up or slowed me down. Because the display tracked with the orientation of my head, whichever way I moved determined not only what I saw but where I went. I pressed the forward button and strolled down the market street and, at Rizzo's instruction, turned at a doorway and entered a house. Inside were two insurgents, one on his knees with his hands tied behind his back, the other dead on the ground. A baby was crying. I moved on.

The next time I put on the headset was in Marina del Rey, Cali-

fornia, at an Institute for Creative Technologies lab space called
FlatWorld, most of which was given over to life-size "mixed reality"
worlds that could be negotiated without special equipment. (It was
so realistic that when a virtual insurgent popped up across the virtual street from the virtual building in which I was standing, his
bullets made successive holes in the virtual wall behind me and
seemed to shower plaster dust through the air.) The Virtual Iraq
design team, two artists and a programmer, worked out of Flat-
World, and it was their system, with the most recent improvements
and additions, that I was using. This time Rizzo sat me in a chair
placed over a bass shaker, which is also known as a tactile trans-
ducer, a device that transmits the feel of sound. I slipped on the display and the headphones, and Rizzo pressed some keys on his computer and made me the driver of a Humvee, with a soldier in desert
fatigues sitting next to me and another in the back. (Because the
gunner was in the turret, when I looked in the rearview mirror I
saw only his boots and his pant legs.) As soon as I started up the vehicle, the floor under me began to vibrate and my ears filled with
the hum of tires on pavement. Suddenly, a gunman appeared on
the overpass above me and started to shoot. Off to my right, a car
burst into flames. Half a second later, the explosion entered my
body through my feet and ears. It was startling, the way any unexpected loud noise is, but it wasn't frightening. Even when the guy
in the seat next to me was shot, and his shirt sprouted a red bloom,
it wasn't frightening. I had never been to Iraq. I had never been to
war. The scene did not conjure any memories for me, traumatic or
otherwise. It was, as JoAnn Difede said of stairs on September 10 to
a person who worked in the World Trade Center, neutral.

I had seen, though, what might happen if it triggered an emotional response when an actor named Ed Aristone, who had been
cast in a movie about the Iraq conflict and wanted to get a sense of
what combat was like, put on the head-mounted display at Flat-
World and found himself in the midst of a war. Rizzo cued up car
bombs, shouting soldiers, ambient city sounds, blinding smoke, inert bloody bodies, the call to prayer, a child running across the
street, the cough of an AK-47, snipers, a nighttime gale — all ten
plagues and their cousins at once. Aristone started to sweat. His
heart was racing. His hands were numb. He was having a hard time
holding the rifle. His face went white. He bit his lips. After ten minutes, he said he'd had enough.

"This shows you why you need a trained therapist," Rizzo said, turning off the machine and watching Aristone, who was bent over with his hands on his knees, taking deep breaths. "Someone who knows exposure therapy, who knows how little things can set people off. You have to understand the patient. You have to know which stimuli to select. You'd never do what I just did — you'd never flood them. You have to know when to ramp up the challenges. Someone comes in, and all they can do is sit in the Humvee, maybe with the sound of wind, and may have to spend a session or two just in that position. For PTSD, it's really intuitive. We provide a lot of options and put them into the hands of the clinician."

One of these is Karen Perlman, a civilian psychologist who uses Virtual Iraq with patients at the Naval Medical Center San Diego. Perlman is an apple-cheeked, middle-aged native Californian with cascading brown hair, who, when I met her, was wearing an elegant short black dress with a pink, blue, and purple tie-dyed silk scarf. At first glance, Perlman does not seem to be the sort of person a young marine would cotton to, but Rizzo says that she has a gift, and so far eight of the nine patients she has treated no longer meet the criteria for PTSD. (This number does not account for those who dropped out.) "It's a very collaborative relationship," she told me in February, when Skip Rizzo and I drove down to San Diego. "I know which stimuli I'm going to add as the therapy progresses. I'm not going to overwhelm them. There are no surprises. I say, 'I think you're ready for the IED blast or for more airplanes.' I'm not only adding more but increasing the duration of each one. It's intensive, but for PTSD you need a treatment that is intensive."

Although Perlman had been a clinician for more than twenty years, before she began work with marines at the Naval Medical Center she had never used prolonged-exposure therapy with patients, and she was surprised by its therapeutic power. (She had spent four days in Philadelphia being trained by Edna Foa, the director of the Center for the Treatment and Study of Anxiety at the University of Pennsylvania, who initially developed the prolonged-exposure technique while treating rape victims, and a day with JoAnn Difede, learning how to integrate virtual reality with exposure therapy.) "I've seen patients recover in five to six weeks," she said. "To see someone respond in such a dramatic way is very gratifying. What we're doing is very structured and systematic. It treats the core fear, the avoidance and the anxiety that are part of PTSD, in a potent way. VR augments the therapeutic process. When the patients start to see results, usually by the fifth session, they turn the corner and get motivated."

Outside his therapist's office, Travis Boyd had "homework." He had been told to listen to an audiotape of the previous session and to do the very things he had been avoiding — going to the mall, driving a car, calling his family back home and telling them what was really going on with him and answering their questions. He also called every one of the men who had been on the roof that day and asked them to tell him their recollections. He was surprised to learn that not one of them thought, as he had for so long, that he was responsible for their team leader's death. In fact, as they remembered it, the man had told Boyd to wait at the top of the stairs. "I had been walking around with all this guilt about getting my brother killed," Boyd said. "It just weighs on you. He was not the only friend I lost, but I was closest to him. Everyone thought it was awful that he died, but nobody thought it was my fault."

The first thing Boyd noticed, after a few weeks of Virtual Iraq exposure therapy, was that he was able to sleep without medication. He was more relaxed, and he could joke around. "Before, I felt like there were two people in me," Boyd said. "The marine, who was numb, who was a tough guy, and the civilian me, the real me, the guy who isn't serious all the time, the guy who can take a joke. By the end of therapy I felt more like one person. Toward the end, it was pretty easy to talk about what had happened over there. We went over all the hot spots in succession. I could talk about it without breaking down. I wasn't holding anything back. I felt like the weight of the world had been lifted. I was ready to be done. The last two sessions, I didn't think I needed to be there anymore."

The last time I talked to Travis Boyd, it was his third wedding anniversary. Boyd is now twenty-two and works for a commercial construction firm in the Midwestern town where he grew up. "Most of the intrusive thoughts have gone away," he said. "You never really get rid of PTSD, but you learn to live with it. I had pictures of my team leader that I couldn't look at for three years. They're up on my wall now."

WALTER ISAACSON

Chain Reaction

FROM *Discover*

IN THE POPULAR IMAGINATION, Albert Einstein is intimately associated with the atom bomb. A few months after the weapon was used against Japan in 1945, *Time* put him on its cover with an explosion mushrooming behind him that had $E = mc^2$ emblazoned on it. In a story overseen by an editor named Whittaker Chambers, the magazine noted with its typical prose from the period: "There will be dimly discernible, to those who are interested in cause & effect in history, the features of a shy, almost saintly, childlike little man with the soft brown eyes, the drooping facial lines of a world-weary hound, and hair like an aurora borealis . . . Albert Einstein did not work directly on the atom bomb. But Einstein was the father of the bomb in two important ways: 1) it was his initiative which started U.S. bomb research; 2) it was his equation ($E = mc^2$) which made the atomic bomb theoretically possible."

Newsweek likewise did a cover on him, with the headline "The Man Who Started It All." This was a perception fostered by the U.S. government. It had released an official history of the atom bomb project that assigned great weight to a letter Einstein had written to President Franklin Roosevelt warning of the destructive potential of an atomic chain reaction.

All of this troubled Einstein. "Had I known that the Germans would not succeed in producing an atomic bomb," he told *Newsweek*, "I never would have lifted a finger." He pointed out, correctly, that he had never actually worked on the bomb project. And he claimed to a Japanese publication, "My participation in the production of the atom bomb consisted in a single act: I signed a letter to President Roosevelt."

Neither the public image nor the personal protests capture the true, complex story of Einstein and the bomb. Contrary to common belief, Einstein knew little about the nuclear-particle physics underlying the bomb. On the other hand, as the archives show, Einstein did not merely sign the letter to Roosevelt. He was deeply involved in writing it, revising it, and deciding how to get it to the president.

The tale begins with Leo Szilard, a charming and slightly eccentric Hungarian physicist who was an old friend of Einstein's. While living in Berlin in the 1920s, they had collaborated on the development of a new type of refrigerator, which they patented but were unable to market successfully. After Szilard fled the Nazis, he made his way to England and then New York, where he worked at Columbia University on ways to create a nuclear chain reaction, an idea he had conceived while waiting at a stoplight in London a few years earlier. When he heard of the discovery of fission using uranium, Szilard realized that that element might be used to produce this phenomenon.

Szilard discussed the possibility with his friend Eugene Wigner, another refugee physicist from Budapest, and they began to worry that the Germans might try to buy up the uranium supplies of the Congo, which was then a colony of Belgium. But how, they asked themselves, could two Hungarian refugees in America find a way to warn the Belgians? Then Szilard recalled that Einstein happened to be friends with that country's Queen Elizabeth.

"We knew Einstein was somewhere on Long Island, but we didn't know precisely where," Szilard recalled. So he phoned Einstein's Princeton, New Jersey, office and was told he was renting the house of a Dr. Moore in the village of Peconic. On Sunday, July 16, 1939, they embarked on their mission with Wigner at the wheel (Szilard, like Einstein, did not drive). But when they arrived, they couldn't find the house, and nobody seemed to know Dr. Moore. Then Szilard saw a young boy standing by the curb. "Do you, by any chance, know where Professor Einstein lives?" he asked. Like most people in town, the boy did know, and he led them up to a cottage near the end of Old Grove Road, where they found Einstein lost in thought.

Sitting at a wooden table on the porch of the sparsely furnished cottage, Szilard explained how an explosive chain reaction could be produced in uranium layered with graphite by the neutrons released from nuclear fission; those neutrons would split more nuclei, and so on. "I never thought of that!" Einstein interjected. He asked a few questions and quickly grasped the implications. Instead of writing to the Belgian queen, Einstein suggested, they should contact a Belgian minister he knew.

Wigner, showing some sensible propriety, suggested that three refugees should not be writing a foreign government about secret security matters without consulting the U.S. State Department. Perhaps, they decided, the proper channel was a letter from Einstein (the only one of them famous enough to be heeded) to the Belgian ambassador, with a cover letter to the State Department. With that plan in mind, Einstein dictated a draft in German. Wigner translated it, gave it to his secretary to be typed, and then sent it to Szilard.

A few days later, a friend arranged for Szilard to talk to Alexander Sachs, an economist at Lehman Brothers and a friend of President Roosevelt's. Showing a bit more savvy than the three theoretical physicists, Sachs insisted that the letter go right to the White House, and he offered to hand-deliver it.

It was the first time Szilard had met Sachs, but he found the bold plan appealing. "It could not do any harm to try this way," he wrote to Einstein. Einstein wrote back asking Szilard to come back out to Peconic so they could revise the letter. By that point Wigner had gone to California for a visit. So Szilard enlisted, as driver and scientific sidekick, another friend from the amazing group of Hungarian refugees who were theoretical physicists, Edward Teller.

Szilard brought with him the original draft from two weeks earlier, but Einstein realized that they were now planning a letter that was far more momentous than one asking Belgian ministers to be careful about Congolese uranium exports. The world's most famous scientist was about to tell the president of the United States that he should begin contemplating a weapon of almost unimaginable impact. "Einstein dictated a letter in German," Szilard recalled, "which Teller took down, and I used this German text as a guide in preparing two drafts of a letter to the president."

According to Teller's notes, Einstein's dictated draft not only raised the question of the Congo's uranium but also explained the possibility of chain reactions, suggested that a new type of bomb could result, and urged the president to set up formal contact with physicists working on this topic. Szilard then prepared and sent back to Einstein a forty-five-line letter and a twenty-five-line version — both dated August 2, 1939 — "and left it up to Einstein to choose which he liked best." Einstein signed them both in a small scrawl.

The scientists still had to figure out who could best get it into the hands of President Roosevelt. Einstein was unsure Sachs could do the job. When Szilard sent back to Einstein the typed versions of the letter, he suggested that they use as their intermediary Charles Lindbergh, whose solo transatlantic flight twelve years earlier had made him a celebrity. All three refugee Jews were apparently unaware that the aviator had been spending time in Germany, had been decorated the year before by Hermann Göring with that nation's medal of honor, and was becoming an isolationist and Roosevelt antagonist.

Einstein had briefly met Lindbergh a few years earlier in New York, so he wrote a note of introduction, which he included when he returned the signed letters to Szilard. "I would like to ask you to do me a favor of receiving my friend Dr. Szilard and think very carefully about what he will tell you," Einstein wrote. "To one who is outside of science the matter he will bring up may seem fantastic. However, you will certainly become convinced that a possibility is presented here which has to be very carefully watched in the public interest."

Lindbergh did not respond, so Szilard wrote him a reminder letter on September 13. Two days later, he realized how clueless he and his colleagues had been when Lindbergh gave a nationwide radio address. It was a clarion call for isolationism. "The destiny of this country does not call for our involvement in European wars," Lindbergh began. Interwoven were hints of his pro-German sympathies and even some anti-Semitic implications about Jewish ownership of the media. "We must ask who owns and influences the newspaper, the news picture, and the radio station," Lindbergh said. "If our people know the truth, our country is not likely to enter the war."

Szilard's next letter to Einstein stated the obvious. "Lindbergh is not our man," he wrote.

The physicists' other hope was Sachs, who had been given the formal letter to Roosevelt that Einstein signed. But Sachs was not able to find the opportunity to deliver it for almost two months.

By then, events had turned what had been an important letter into an urgent one. At the end of August 1939, the Nazis and Soviets stunned the world by signing a war-alliance pact and proceeding to carve up Poland. That prompted Britain and France to declare war.

Szilard went to see Sachs in late September and was horrified to discover that he still had not been able to schedule an appointment with Roosevelt. "There is a distinct possibility Sachs will be of no use to us," Szilard wrote to Einstein. "Wigner and I have decided to accord him ten days' grace." Sachs barely made the deadline. On the afternoon of Wednesday, October 11, he was ushered into the Oval Office carrying Einstein's letter, Szilard's memo, and an eight-hundred-word summary he had written on his own.

The president greeted him jovially: "Alex, what are you up to?"

Sachs worried that if he simply left Einstein's letter and the other papers with Roosevelt, they might be glanced at and then pushed aside. The only reliable way to deliver them, he decided, was to read them aloud. Standing in front of the president's desk, he read his summation of Einstein's letter and parts of Szilard's memo.

"Alex, what you are after is to see that the Nazis don't blow us up," the president said.

"Precisely," Sachs replied.

"This requires action," Roosevelt declared to his assistant.

The following week, Einstein received a polite and formal thank-you letter from the president. "I have convened a board," Roosevelt wrote, "to thoroughly investigate the possibilities of your suggestion regarding the element of uranium." Still, the effort's slow pace and meager funding prompted Szilard and Einstein to compose a second letter urging the president to consider whether the American work was proceeding quickly enough.

Despite helping to spur Roosevelt into action, Einstein never worked directly on the bomb project. J. Edgar Hoover, the director of the FBI even back then, wrote a letter to General Sherman Miles, who initially organized the efforts, that described Einstein's pacifist activities and suggested that he was a security risk. In the end, Einstein played only a small role in the Manhattan Project. He was asked by Vannevar Bush, one of the project's scientific overseers, to help on a specific problem involving the separation of isotopes that shared chemical traits. Einstein was happy to comply. Drawing on his old expertise in osmosis and diffusion, he worked for two days on a process of gaseous diffusion in which uranium was converted into a gas and forced through filters.

The scientists who received Einstein's report were impressed, and they discussed it with Bush. In order for Einstein to be more useful, they said, he should be given more information about how the isotope separation fit in with other parts of the bomb-making challenge. Bush refused. He knew that Einstein didn't have and couldn't get the necessary security clearance. "I wish very much that I could place the whole thing before him and take him fully into confidence," Bush wrote, "but this is utterly impossible in view of the attitude of people here in Washington who have studied his whole history."

Thus the scientist who had explained the need for a bomb-making project was considered too risky to be told about it.

FREDERICK KAUFMAN

Wasteland

FROM *Harper's Magazine*

IN 1998 John Brunston bought a house in Yorktown Heights, a suburb of New York City. His wife decorated bedrooms for their three daughters, and Brunston planted a cherry tree and an American flag. Then came the spring rains, and Brunston discovered a dark fountain of human waste bubbling up from his back yard.

He did everything he could to stop the sickening flow. He consulted engineers, installed a new septic tank, purchased sump pumps, dumped ton after ton of fresh soil over the ooze. He spent tens of thousands of dollars, but the evil-smelling gloop still percolated to the surface.

Brunston lives in a densely populated, well-established neighborhood. He should not have to use a septic tank. His waste should flow into the underground pipe that lies no more than thirty feet from his front door, and that pipe should carry the Brunston family waste far from the Brunston family home. But neither John Brunston nor anyone else on this lovely block of Yorktown Heights can hook up to a sewer, because the sewers of Yorktown Heights are already full. In fact, they are running at 100,000 gallons of waste per day beyond capacity. So Brunston's back yard must absorb Brunston's waste. And it cannot.

Every day, America must find a place to park 5 billion gallons of human waste, and our country appears increasingly unable to find the space. Not surprisingly, the effects have been dramatic: the *Colorado Springs Gazette* reported that one Jennifer McCowen discovered a geyser of raw sewage emerging from her toilet. "I couldn't believe it," McCowen told the newspaper. "It filled the bathtub un-

til it overflowed." In southern California, where surfer Web sites
post hourly runoff warnings, a paltry 2-million-gallon belch will not
stop a dude from his appointed rounds in the bays of Santa Monica
or Hermosa Beach. But when an aging main in Oahu discharged
48 million gallons of human waste into the placid waters of Waikiki,
residents were not happy — particularly not the one who fell head-
long into the fetid morass and died. In Durham, North Carolina,
sewage has reared up from the depths and gurgled across the city
sidewalks at the alarming rate of once every eleven days. North
Carolina has notched more than two thousand such spills, both urban and suburban, and the state of Oregon fined Portland a half-
million dollars for sixty-seven overflows. Local newspapers from
Tulsa to Allentown describe the same nightmare: reeking goo in-
vades family basement and living room. Unclear who will pay for
the mess.

   This sounds like a problem. For thousands of years, *Homo sapiens*
flocked across continents in pursuit of bird, beast, and fresh water,
leaving behind a trail of gnawed bones and steaming waste. The
moment we stopped removing ourselves from that waste, it had to
be removed from us. Thus the origins of civilization; thus the glories of Rome, Paris, and Philadelphia; thus the horror of John
Brunston's back yard. A civilization that cannot escape its own fecal
matter is a civilization in trouble — unless, of course, the uneasy relationship between man and his effluents can evolve. Perhaps we
could bridge the chasm, heal the rift, transform the untouchable
into something rich and strange and marketable. Or so I hoped as I
toured John Brunston's back yard.

   The soggy lawn squished beneath our shoes, and I surveyed the
wet grass with suspicion and growing anxiety. We smushed beyond
the cherry tree. I kneeled in the shade, tugged at a tuft of grass,
and the earth peeled back like a scab, releasing the dreaded stink. I
recalled a report from the National Research Council, entitled
Biosolids Applied to Land, in which the authors noted that "odor perception has been shown to affect mood, including levels of tension, depression, anger, fatigue, and confusion." Brunston smiled
vaguely and asked that I not include his real last name in the article. He told me he was embarrassed about his house.

   I poked a stick into the lawn. A slick of clay lurked beneath the
soft soil, then a smattering of damp stones, then the terrible stew. I

turned away gasping, profoundly sorry I had made the trip. Since
infancy we have been taught to stifle our curiosity, programmed
not to look. Perhaps there were good reasons for the repression
and denial.

"Want to see the septic tank?" asked Brunston. We made our way
past the lilac bushes and the bird feeder, then he pushed aside a cedar rocking chair, removed four paving stones from the patio, and
set about unscrewing the white plastic tank cover. Instead of watching him, I gazed at the nearest pine tree. A rope hung from a
branch, and a well-battered baseball hung from the rope.

"There," he said. "It's full. Can't digest any more."

The white van speeded through miles of concrete tunnel. "The first
regulations with respect to waste go back to the Code of Ham-
murabi," said Steve Askew, superintendent of New York's North
River Wastewater Treatment Plant, one of the world's largest. "You
have to bury your waste far from where you sleep." And he gave me
the look. Steve Askew never finished college, but that look had
seen to the bottom of things. It was both spooky and intimidating,
that particular look of pity and loathing the wise bestow upon the
ignorant. He knew something I wanted to know: the ultimate fate
of our waste.

"People wake up in the morning, they brush their teeth, flush
the toilet," said Askew. "They think it goes to the center of the
Earth."

If you happen to live within one particular 5,100-acre patch of
the West Side of Manhattan, instead of going to the center of the
Earth, your waste flows to Askew's extraordinary concrete cesspit:
twenty-eight concrete acres suspended above more than two thousand concrete caissons sunk into the shallows between the West
Side Highway and the Hudson River. Constructed in the 1970s,
topped by three swimming pools, a skating rink, and a carousel,
North River cost the city a billion dollars, 100 million of which went
straight into odor control.

North River is just one of New York City's fourteen wastewater
treatment plants, the first of which opened in 1886, along with the
Statue of Liberty. These plants handle every conceivable kind of
sewerable waste from the city's 8 million permanent residents, not
to mention anything a commuter or a tourist might care to add.

They separate the material that comes their way into solid, liquid, and gaseous parts, which they further subdivide into that which must be discarded, that which may be consumed, and that which someone, somewhere, might eventually be able to sell.

The substance that enters North River is mostly water, and the vast majority of that water leaves the plant after not much more than six hours, disinfected to the extent that it can merge inoffensively with the Hudson River. One flush on the Upper West Side at seven in the morning, and by three in the afternoon the water is back on the street, so to speak. What's left over is a half-million gallons of concentrated daily waste, now known as sludge.

The white van had reached the end of its journey, and I followed Askew into an enormous room of computers, controls, workstations, and switches. Behind us flashed a wall-size diagrammatic panel, the great computerized brain of waste. Next to us stood the oiler, who had been at North River twenty years.

"Right now we're at 135 million gallons per day," said the oiler.

The greatest increase occurs between eight and nine in the morning, when the city's output swells from 70 million to 150 million gallons per day. This is known as the big flush. Now it was 11:00 A.M., and in a few hours the circadian flow of biology en masse would begin to diminish, eventually bottoming out around four in the morning, at 68 million gallons per day. The rhythm is as steady as the tides. "The Super Bowl halftime surge is a myth," said Askew.

He led me across the concrete floor, through a concrete warehouse, and to the concrete screening room, where he began to extol the virtue and beauty of his eleven-mile-long sewage interceptor. By the time the morning flush finally rolls into North River, it has joined the downstream flow of all the other morning flushes from all the other sewage lines from Bank Street to the Upper West Side and has sunk fifty-four feet below sea level. It is here, at the extreme low point of this immense underground current, that North River gets to work. In the stygian depths, its mighty diameter swollen to sixteen feet, the dark torrent branches into six channels, each of which must be pumped to the top floor of the plant, where gravity can once again take hold and set the outcast on a new journey.

Askew gazed into the inky pool of untreated wastewater and began to describe some of the marvels the interceptor had disclosed.

Aside from the daily take of leaves, sticks, cans, and paper, the great rake had brought up quite a few vials of cocaine. When cops bang on the door, the toilet is a drug dealer's best friend. Ditto for the professional forger: a good deal of counterfeit money has floated into Steve Askew's hands. Twenty years ago a dog showed up, a living dog that became the mascot of a Brooklyn plant.

"I never saw an alligator," said Askew.

As we walked away from the pool, I asked about the wind. No matter what the weather was outside, no matter where we traveled inside, the thick concrete walls of North River generated bracing gusts. Askew explained that every minute, titanic blowing machines inhaled 600,000 cubic feet of fresh air and exhaled 750,000 cubic feet of carbon-filtered, bleach-scrubbed exhaust — six to twelve complete air changes per hour.

But the scouring of North River's halitosis, while essential to community relations, has nothing to do with the plant's core mission. The alchemy of purgative transformation starts in the warmth and humidity of the next chamber we visited, where submerged chemical mixers combine the waste with custom-made bacteria. "It's volatizing off!" Askew yelled above the din of engines and bubbling brown water. Undeterred by the general uproar, Askew detailed the technical intricacies of fecal breakdown and development, but I'm afraid the cacophony blunted the nuances. So Askew dumbed down the lecture. "This looks really good!" he hollered. "Tan water! Light brown froth! Small bubbles! Musty smell! If the foam looks like chocolate mousse, that's an indication of a bacteriological process!"

We headed to a low-ceilinged room so huge it did not appear to have walls. Here were the settling tanks, the final stop before the water returned to the world. Peace held sway among these last lagoons, and indistinct reservoirs misted into a concrete vanishing point hundreds of yards away. "On a cold morning, you will see the water vaporing off," Askew said. "And it will rain inside the plant."

He gave me the look. "When it is really cold, it snows inside the plant."

At that moment, two square football fields of submerged jets spumed into the shadows, and the bronze liquid arced, more sublime and terrifying than the fountains of Trevi or Versailles. Soon these waters would sluice down concrete courses to mix with the mighty Hudson. As for the remaining sludge, it also would depart, but by an altogether different route.

When the froth finally settled back into silence, Steve Askew backtracked through the concrete dungeons until we arrived at a perfectly normal conference room and a nice surprise — someone had ordered pizza!

Despite the skating rink and swimming pool, despite the bleach, the carbon filters, the white hardhats and the spotless lab coats of the technicians, despite the banks of UNIX computers and the sober talk of asymptotes and oxygen demand, despite the board-room-size wood-veneer table and the well-upholstered ergonomic chairs and the rush of 20,000 cubic feet of air per second, and despite, to put it bluntly, one of the most extraordinary concealments in all of human history, North River still managed to evoke unappetizing associations. But as I gazed at the cheese and red sauce and blackened crust, I recalled the words of one of the many wastewater professionals I had met that morning: "One of the things about the job — you still have to eat."

So I sat down to lunch and learned about the glorious future of waste. Now that biochemists could scour the particles on the atomic level, the plant could recover ibuprofen, acetaminophen, endocrine disrupters, DEET, Prozac, and Chanel No. 5. Even caffeine could be extracted from the mix, and I had a hunch the citizens of New York excreted boatloads of stimulant. Perhaps Starbucks would be interested. The technology was there.

"Twenty years from now we will be removing things we have no idea about," said Askew. "Penicillin, mercury, heroin. Will this be a pharm business? An energy business? An agribusiness?"

He took another bite and delivered the look.

"A bear goes in the woods and it takes two years to decompose. We do it in six hours. In six hours, we imitate all of nature — from the big bang to the big chill. We're trying to put it back the way that God intended."

Throughout its long history of denial, waste has lurked behind countless appellations: egesta, dejecta, sharn, stale, skite, dynga, ordure, oriental sulfur, occidental sulfur, and carbon humanum, to name but a few. Witches' potions called for *etihs;* alchemists' elixirs required *botryon, aureum, oletum,* or *zibethum.*

The rich and variegated literature of waste has suffered the same repression as the language. *The Secrets of Physicke* appeared in London in 1633, and enumerated all uses of pedung, but nothing more was heard upon the subject until more than half a century later, when a Frankfurt publisher issued Christian Franz Paullini's scandalous *Dreck Apothek,* in which the esteemed German botanist guaranteed a successful cure for "even the most difficult, most poisonous diseases and bewitched injuries from head to feet, inside and out, with filth and urine." *Dreck* became the undisputed authority on all stool-related matters for almost three decades, until a Dresden publisher brought out M. Schurig's vast and ponderous *Chylologia* (only five short years after publication of his equally vast and ponderous *Spermatologia*). *Chylologia* contained citations from nearly seven hundred scholars of human excrescence, each more unknown than the next: Sclopetarius, Goclenius, Spagyria Microcosmi, and Zacutus Lusitanus. To write about the subject guaranteed obscurity.

The sole American to contribute to the literature was Captain John Bourke of the United States Army. After Bourke fought in the Civil War, he traveled west to fight the Apaches and, generally speaking, keep any and all natives in line. Bourke got hooked on waste after he witnessed a Zuni urine dance in which a great *olla* of urine provided a "strange and abominable refreshment." He must have spent countless solitary hours thereafter in the fort library perusing the secret ingredients of ancient sterility cures and primitive love philters. After a decade of scholarship, he emerged with the most thorough study of excrement ever published in this country, *Scatalogic Rites of All Nations.* Bourke described the fecal practices of the Hualapai and the Navajo, the Tartars and the Fiji Islanders, the Egyptians and the Hottentots, the Samoans and the Bongos of the upper Nile. His sources ranged from Martin Luther to Montaigne, Moses, Martial, Marco Polo, Ezekiel, Erasmus, and Shakespeare. Upon publication of his work in 1891, Bourke's editors stamped the frontispiece "Not for General Perusal."

The North River Wastewater Treatment Plant creates sparkling fresh, nutrient-rich sludge. In the old days, the night soil collector would spread such promising young shite on village crops, but in these days of refinement and paranoia, sludge requires a few more alchemical interventions and changes of venue. Thus did I find myself on the bridge of a sludge boat next to a potbellied man who swiveled his chair and checked the radar.

"We're the secret," said Captain Jonas.

Captain Jonas is from Flushing, and he was quick to tell me about the story he had read the other day that said New York City has the fourth-largest navy in the world. "For a municipality to have its own fleet of tankers is virtually unheard of," he said. Twenty-four hours a day, seven days a week, the New York City navy moves New York City waste.

The boat plowed past the United Nations, and not one diplomat who gazed out the window could have suspected his part in our journey to the heart of darkness. Half a nautical mile below Wall Street we would hang a right, then head up to Harlem's western shore. Our mission: to pick up a single load of 700,000 gallons of waste from North River and carry it to Wards Island in the Bronx, where the sludge could be transformed into cake. Melville once wrote that a whaling ship had been his Harvard and his Yale, which made me consider the pedagogical value of our prospective cargo.

The sludge boat stretched longer than a football field and packed two massive propellers, two titanic cranes, 3,000 horsepower, and one frighteningly distended black rubber Goodyear hose, the diameter of which matched the length of my leg. We all knew what would go through that hose.

Captain Jonas pulled back the throttle, and the sludge boat quaked, lurched, and churned past the Statue of Liberty. "It may be shit to you," he said. "It's bread and butter to me."

I asked him about the future of human waste. "You can divert it, but you can't stop it," he said. "It's a problem now, it'll be a problem in the future."

We bulled past Governors Island, which is rumored to have no sewage system. I asked Captain Jonas what happened to Governors Island waste.

"It drops into the Upper Bay," said Jonas.

"Straight into the water?" I asked.

Now Captain Jonas gave me the look. "The ban on ocean dumping was rammed down people's throats," he said. "Ocean dumping was not the big, monstrous evil."

"It was the *best* fishing ground," lamented the first mate.

"Fifty years from now we'll probably be ocean dumping again," said Jonas. "This is cyclical."

The great concrete mass of North River loomed ahead, and plant staff donned their hard hats and work gloves and readied themselves for pumping. Captain Jonas marched outside. "Right twenty!" he roared to the first mate. "Right twenty! Back ten!"

One of the onboard cranes lifted the mighty black snake, and the plant workers grabbed the hose, wrestled it into position, and gave the ready signal. Someone turned a valve. The hose jumped and twitched, the ship trembled with the force, and a sour smell began to rise. "Gotta go!" cried Captain Jonas. "Gotta go! Gotta go!"

As the boat filled with waste, I descended the stairway to the main deck. One of the hardhats was embracing the rank and monstrous intestine, trying to hold the writhing rubber steady, and I gave him plenty of leeway. But after a while I stepped toward the hose, touched my palm to the warmth, and felt the cosmic surge.

The veneration of human waste boasts a noble history. Amid the verdant passes of the Himalayas, intrepid Jesuit missionaries discovered cult worship of multicolored powders and hand-fashioned pills produced from the dried and pulverized ejecta of the Grand Lama, which Buddhists wore as amulets around their necks. Some Mongols painstakingly packed the holy relics within golden boxes. Others consumed it as sacred snuff, still others as a rare condiment. "When they feast their friends," noted one witness, they "strew it upon their meat."

In the *Scatalogic Rites of All Nations*, Captain John Bourke wrote that in the distant past, "all excretions, solid or fluid, were invested with mystic properties," an assertion that might go far in explaining why the creation myths of the Australian aborigines avowed that the great Bund-jil filled the oceans with his urine and the obscure deity Mingarope molded men and women from her feces. Of course, modern religion has long sought to expunge the ancient gods and goddesses who touched and ate and even loved ejecta. Consider Saturn, most ancient of the Roman gods, who was also known by the epithet Sterculius — as in *stercus*, or dung — god of the magical transformation of death into life. The great spirit of this original, unexpurgated Saturn inhabited manure. *Aurum de stercore.*

What could be more magical, more *godlike,* than the metamorphosis of that which we abhor and expel into that which we desire, embrace, and ingest? On the far eastern peninsula of Russia lie the snow-blanketed mountains, shooting geysers, and hot springs of Kamchatka. Like many isolated folk, the Kamchatkans retained their own particular worldview far past the time when the "primitive" had been drummed out of most of Asia and Europe. Of all the Kamchatkan deities, Kutka was the greatest. Kutka created the world and every living being — then fell in love with his excrement and wooed it as his bride.

Black-magic hexes could be undone only by the potent charms of human waste. The exorcism rites of the Abyssinians demanded waste, as did the oblations of the Ojibwa and the Huron, the Iroquois and the Eskimo, the Mojave and the Patagonians. Alloyed with musk and ambergris and set smoldering, the acrid smell would have been recognized across barbaric Europe as holy.

The most worshipped and praised of all ancient sewers was Rome's Cloaca Maxima, whose spirit resided within the shrine of the goddess Cloacina, where warriors came to purge themselves after battle and young couples purified themselves before marriage. The lovely Cloacina was an emanation of Venus, and her statue overlooked the imperial city's sewer pipes as they transported 100,000 pounds of ancient *excrementum* a day. Built in the sixth century B.C. by the two Tarquins, hailed as one of the three marvels of Rome, the Cloaca became one of the city's great tourist traps. Agrippa rode a boat through it. Nero washed his hands in it. "Thus may the greatness of Rome be inferred," declared Cassiodorus. "What other city can compare with her in her heights, when her depths are so incomparable?"

The ancient doctors Hippocrates, Xenocrates, and Dioscorides employed plasters and poultices and styptics and decoctions of waste to treat holy diseases, such as epilepsy and mania. Physicians paid close attention to filth remedies for the plague and boils, headache and insomnia, dementia and insanity, not to mention anorexia, cancer, cataracts, convulsions, constipation, and freckles. Proto-psychotherapists analyzed *melancholy excrements.*

Like the Romans and the Moabites and lovesick maidens in France, the alchemists of Europe believed in the spiritual powers of human waste, which ranked among the strongest of all magnetic medicines. The great Paracelsus, father of modern pharmacology, kept a store from which he hoped to conjure nothing less than the philosopher's stone. "Man's dung, or excrement, hath very great virtues," he wrote, "because it contains in it all the noble essences."

Here are the ABCs of Frederick Kaufman's waste: it flows into a sewage pipe, ferments and settles at North River, then splashes through a black rubber hose. It takes a trip around the southern tip of Manhattan and up to the Bronx in Captain Jonas's tanker, disembarks on Wards Island, then falls into a giant blue centrifuge, where it begins to spin, faster and faster, until it has become an incomprehensible blur at 2,000 revolutions per minute.

I stood on the floor amid the roar of thirteen German-made Humboldts and watched as Joe Pace, a twenty-one-year Department of Environmental Protection veteran, turned one of them off and leaned down to scoop out a few tablespoons of black, carbony dust. The liquid sludge from North River had been dried and shrunk into fine gravel, and the smell was amazing. Overcome the repression, I told myself. Transcend it. Do not mind the pain.

Joe shone a flashlight on the cake, and it was blacker than black, the stink concentrated ten or fifteen times normal. Tears dripped down my cheeks. "It takes a while to get used to it," said Joe.

He flipped the switch, and as the centrifuge started back up I touched the blue cast iron and felt the vibrations of the orbit, felt the warmth. Joe kept his cake at body temperature. He turned an orange knob, and clear fluid poured out the bottom of the tank and pooled on the floor, not too far from my loafers. He shone his flashlight on the discharge. "Not bad," he said. "Pretty clean."

A dark conveyor belt shuttled the black dust from the blue centrifuges to a weigh station. Every ounce of material had to be counted before it could be released to the hoppers. Somewhere, someone was keeping score. Joe led me to the giant closed garage, where we watched a fresh load drop into the back of an eighteen-wheeler.

"They carry twenty-five tons," said Joe.

It did not take very long before the truck's hold was filled, at which point a black hood automatically unfurled across the top. The truck pulled out, the eighteenth that day to rumble from the Wards Island dewatering plant and into the streets. No one would have suspected the nature of its contents.

I asked who ran the trucks.

"NYOFCO," said Joe.

We left the garage and headed to the control room, where Joe explained the red and yellow lights of his mimic board, and I nosed around until I found an old piece of stained paper Scotch-taped to the wall. NYOFCO, it said at the top.

Joe checked one of his screens and picked up the phone. "Be here in an hour," he told me.

"Who will be here?"

"NYOFCO," said Joe.

He turned back to the flashing mimic board and I to the stained memo on the wall. NYOFCO was the New York Organic Fertilizer Company, which was itself a wholly owned subsidiary of another company, called Synagro. And what was Synagro? In 2005 Synagro Technologies, Inc., somehow managed to sell nearly half a million tons of human waste for revenues of $338 million. The company did its business on a million acres of land in thirty-seven states and had signed deals with six hundred colossal collectors for the drying, composting, incineration, and product marketing of human waste. The New York Organic Fertilizer Company was just one of many Synagro subsidiaries.

I asked Joe what NYOFCO did with my waste after they hauled it from Wards Island.

"They sell it to the Arabs," he said. "It works in the desert."

When I first set off on the trail of waste from the sewer to wherever it finally ended up, no matter how close or far away, I had believed I was Kurtz heading into the heart of darkness; but somewhere along the way I had morphed into Woodward and Bernstein, following the money.

As it turns out, the transformation of human waste into articles of commerce dates back to ancient Egypt. In the New World, Hernán Cortés reported that human excrement was collected in Aztec sludge rafts, then sold in the marketplace of Tenochtitlán. A seventeenth-century physician named Rosinus Lentilius recounted that the Chinese and Javanese exchanged human waste for tobacco and nuts.

Contrary to its name, waste can be useful. Human tyrd has polished gold, bleached wool, and helped produce salt and cheese. Innumerable tradesmen have used it to tan leather, adulterate opium, eradicate dandruff, ink tattoos, promote hair growth, and brush their teeth. Much to the delight of professional bakers, the *General Homœopathic Journal* in 1886 reported that "chemists have evidently no difficulty in demonstrating that water impregnated with 'extract of water-closet,' has the peculiar property of causing dough to rise particularly fine."

In the early 1980s, the Environmental Protection Agency began a campaign to acclimate U.S. consumers to the commercial use of human waste. Touted as superior to cow manure and commercial fertilizers, ton after ton of EPA-subsidized sludge and cake arrived in low-income rural areas, distributed free of charge to cash-strapped farmers. The EPA knew the ocean-dumping days would soon be gone, knew that the future of waste would lie much closer to home. In fact, the Ocean Disposal Ban Act of 1988 specifically barred human waste from the sea, which meant a sudden need for more holding tanks, more solid-waste treatment plants, more monstrous black hoses, more blue Humboldt centrifuges, and entirely new industries.

Synagro Technologies calls itself the largest recycler of biosolids in America, the only national corporation focused exclusively on what has come to be known as the "organic residuals industry," a market Synagro hopes will generate $8 billion annually. But it turns out that as the waste revolves, commercial opportunities abound. You may have given it away, but if you want it back you'll have to pay for Granulite, Milorganite, Soil Rich, Vital Cycle, or many of the other wonder soils available from Agway, Home Depot, Kmart, Target, Wal-Mart, or your local purveyor of organic fertilizer. And that's just the beginning. Some of the new breed of waste and pollution entrepreneurs do possess rather sinister names — Controlotron, MicroSepTec, Toxalert, and SICK, Inc. — or even terrifying names, such as American Pulverizer and Annihilator. Not to worry. You can trust your investment in these little guys, because the big guys are in on it, too: Dow, Honeywell, Monsanto, Siemens, and Toshiba.

Now consider that the twenty-first-century waste-management purchase order will have to enumerate innumerable widgets, from mist eliminators to ozonators and vortex meters. How else to defoam, degrease, degrit, demineralize, desalinate, and deionize? We will have to purchase rotocages and rotoscoops, grit chambers and chopper pumps, microbubbles and floating sludge blankets. We will need plastic tubs of next year's coliforms and designer slimes, flocs, fungi, high-tech bacilli, and superdeluxe electrochemical bacteria ready to power tomorrow's superdeluxe microbial fuel cells.

The economic potential of human waste has driven new research agendas for toxicogenomics, odor streams, and vapor media. An industrial chemistry of coagulation and flume has arisen, a new biology of micronutrients, microspores, putrescible organics, and "the human receptor," which is waste-speak for you and me. Will waste become an agribusiness? A biopolymer business? A pharm business? An energy business? Once everyone gets over the mental hump of cadaverine and putrescine, why not consider futures and options trading on the transcontinental waste exchange?

Just as the esoteric mysteries of Bund-jil, Saturn, and Cloacina have been vanquished by biocriteria, particle indices, and risk-management flow charts, so waste itself has been monetized into *anthropogenic input* and the *allochthonous organic matter source*. The all-waste society masquerades as the zero-waste society, an antiseptic land where death itself has been transformed into biocide.

I stopped at the guard gate of the New York Organic Fertilizer Company, then pulled into a parking spot a yard or two from the murky waters of the East River. I got out of the car and took a deep breath. In order to grasp the future of waste, I needed to get past the middle managers I was about to meet here in the Bronx. In the next few hours I would have to make fast friends, then talk my way into an appointment with the big guy in Houston, Synagro's chief executive officer. And that was far from a done deal.

A barbershop-striped smokestack towered above NYOFCO's seven-acre property, and railroad tracks ran close by eight tremendous human-waste storage silos, each capable of holding half of New York City's daily concentrated output. Before I pushed through the glass doors of the main building, I stopped to consider the corporate logo. Beneath smooth swirls and abstract arcs worthy of AT&T or Chevron glowed the Synagro motto: "A Residuals Management Company."

I walked into the lobby and the now-familiar stink of rotten cabbage, dead mouse, and feces. I stepped across the old linoleum to examine a wood-paneled wall plastered with the covenants that allowed NYOFCO to do what it did: chemical-bulk-storage certificates, sewer-use regulation amendments, wastewater-discharge authorizations, and a permit for a 10,000-gallon tank of sulfuric acid. I met John Kopec, fifty-nine, army veteran and Yankees fan. Before he began at NYOFCO, Kopec worked at a cement plant for twenty-nine years.

"A process is a process," said Kopec. "Both have to do with heat."

As usual, the first stop on the visit was a plush chair in the conference room. A cut-glass ashtray sat between us, a Deer Park water cooler stood off to the side. Behind us the shelves groaned with white plastic loose-leaf notebooks, rubber gloves, and gas masks. John Kopec pulled out some samples of Granulite, the product he manufactured from New York City sludge. He asked me to open a vial of the pellets, which I did. He asked me to smell it. I did. Then he told me that the New York Organic Fertilizer Company could produce up to 2,100 tons of these kibble-size nutrient-rich human fertilizer pellets each week.

I nodded as Kopec reviewed all that had happened at North River and at Wards Island, and I smiled as he explained his own system of pin mixers, rotary dryers, purge cycling, and 1,000-degree waste-baking ovens. Kopec described air streams and gas streams, nodalization and cyclones, separators and regenerative thermal oxidizers, and the tale reached its climax when out of the last screener dropped the pellet I now held in my hand.

"The final product is pathogen free," said Kopec, by which he meant no traces of meningitis, hepatitis, or malignant protozoa. No tapeworms, no whipworms, no oocysts, and no streptococci. "There's nothing in this material," said Kopec.

The first part of our interview had come to an end, and a tour of the storage silos was next. "Want to go inside?" he asked, and gave me the long-expected, long-awaited look.

I stood up, put on a hardhat, and adjusted the plastic goggles. We left the building and walked across the dirt straight into Silo 5. We stood in the middle of the great steel cylinder and stared at the steel walls, craning our necks to examine the steel ceiling. The vast silo was empty, washed, and sparkling clean, but its penetrating ammonia scent held us in thrall, as though the waste had permeated the steel. We reeled to the sweet-smelling Bronx outdoors and in-

spected some railroad freight cars, each of which could hold one
hundred tons. Next to the tracks stood a ten-foot tank of frozen nitrogen that pumped gas inside the silos to keep the atmosphere inert. Waste is the stuff munitions are made of, every pile a potential
explosion. The first World Trade Center attack used a fertilizer
bomb.

A sixteen-wheel NYOFCO truck of incoming now approached
Kopec's plant, and we walked over to watch its reception. Of
course, we could not see the freight, shrouded as it was beneath
tightly battened tarps. An automatic door rose as the truck hissed
and beeped its way inside the dark garage; then the door descended and Kopec explained that at that moment, out of sight,
product was streaming into his plant. Then the door rose, and,
dripping from its hose-down, the empty truck emerged and rum-
bled off. The process had taken two minutes.

As we headed back to Kopec's office, I complimented him on his
spotless grounds, and he beamed. A bright red apple shone on his
desk, the fruit perfectly aligned with his cup of Starbucks and the
pack of Utz potato chips. Pictures of the family stood between shiny
model trucks and a box of Kleenex. Kopec explained that a new
duct system was on its way, which meant the interior of the building
would in a short time be untainted by the slightest aroma. It occurred to me the guy was a clean freak.

Where did the pellets go when they left NYOFCO?

Kopec explained that after the railroad cars pulled up beneath
the silos and the dehydrated buckshot tumbled down, the waste
headed to Florida to fertilize our morning orange juice. But that
was not all. Kopec described the Lehigh Cement Company's Maryland plant, which used some of Kopec's pellets — *my pellets* — for
fuel. Burning waste did create a fair bit of ash, so Lehigh dumped
the detritus of the detritus into the concrete mix. The foundation
for tomorrow's skyscraper.

"This is an amazing industry," said Kopec. "We're still in infancy.
We're exploring possibilities."

The conversation had reached the usual endpoint, the dreamy
future of waste. As if we just sat there long enough and thought
about it, we would know. Would pellets heat houses? Light buildings? Fuel cars?

"They're actually exploring it," said Kopec. "Anything is possible."

Finally, I told him about my wish to visit Synagro's world headquarters in Houston. I told him I wanted to interview the CEO. I
knew his name was Robert Boucher and he was forty-one years old,
but that was it.

"He's accessible," said John Kopec. "If I need to talk to him, I can
talk to him."

I looked at the phone.

I had vowed not to stop until I reached the end of the line and saw
the circle close, the beginning in the end, but the closer I came to
the redemptive moment, the more I came to realize that not everyone had been convinced of the miraculous future of human waste.
From the Bronx to Temescal Canyon, complaints have arisen about
a grave threat to nature and humanity. The citizens of Kern
County, California, fought to ban the dumping of human waste,
which the local Green Acres farm had eagerly adopted to fertilize
the wheat, alfalfa, and corn they sold as feed to nearby dairy farms.
No one in Kern County wanted to drink the milk that flowed from
cows that ate the feed that grew from pellets that arrived from
NYOFCO that came from Wards Island that emerged from North
River. That had come from me.

Human waste is, of course, one of the oldest fertilizers known to
human beings. But the future of waste must take into account a few
salient facts: you and your next-door neighbor may be hooked up
to a sewer, but so are DuPont, Monsanto, 3M, and your local hospital, which can make for some far-ranging effluvial consequences.
Content surveys have uncovered dioxins, furans, and coplanar po-
lychlorinated biphenyls, not to mention the germs of pneumonia
and encephalitis. In 1993 the EPA assessed 126 "priority pollutants"
in solid waste. Arsenic, lead, and mercury led the way.

And so the reported incidents of disease near fields of waste and
the growing roster of men, women, and children suffering from
blisters, boils, nose scabs, pleurisy, and fungus in the lungs. So the
reports of hundreds of cows wasting away and dying on farms outside of Augusta, Georgia — cows fed from hayfields fertilized with
the sewage of the residents of Augusta, Georgia. So the opposition
to composting and pellet farming from such groups as the National Sludge Alliance and Citizens Against Toxic Sludge. So the
Sludgewatch e-mail Listserv.

Every year America processes more than 5 million dry tons of sewage sludge. Much of it is slingshot into forests or injected beneath the surface of the Earth, and the remainder fills strip mines and gravel pits or rumbles off to turf farms, aquafarms, tree nurseries, or state parks. Some of the waste creates the rolling greens of your local golf course, the fresh soil of your cemetery, the fertilizer on your front lawn.

Cows graze on treated pasture, as well as on field corn and sweet corn grown on human waste. And when a cow eats pasture, she also eats the dirt from which it springs. Sheep may ingest up to a third of their diet as straight dirt. Then we eat the meat.

The best and the brightest and the most intensively treated human waste meets EPA criteria for "exceptional quality." EQ waste may be used to nurture such human-food-chain crops as beans, carrots, melons, potatoes, and squash. Heinz and Del Monte have taken a cautious approach and decided not to accept ingredients grown on land treated with biosolids. Then again, some organic farmers have reported that human fertilizer raised the protein content of their wheat.

It was eighty degrees outside and pouring rain as I took JFK Boulevard south to the beltway and headed toward the Galleria. I drove through the deluge until I came to Capital One and Texas American Title and the Bering Drive Church of Christ. I pulled into a concrete garage and approached a sinister monolith of a building, the top floor shrouded in mist. This was Synagro, home to the future of waste, and the tinted curtain wall reflected storm clouds.

I stood in the silent, spotless lobby and admired the stainless-steel columns and the black granite floors. Absolutely no one else was there. I caught an elevator to the top and pushed through the double glass doors.

On the walls, Synagro displayed beautiful images of its product. One glossy color photograph featured the barbershop chimney of NYOFCO; another particularly strong composition centered a quaint red barn behind a massive Synagro composter. A framed certificate announced that *Houston Business Journal* had declared Synagro number 93 among Houston's "Top 100 Public Companies."

A secretary led me past a bowl of mints to the boardroom and told me to wait for Mr. Boucher. The conference table here was bigger than any other I had witnessed in my travels through wasteland, and a wooden podium branded with the Synagro logo stood at one end of the long room, flanked by an American flag and a Texas flag. A large oil painting of seven horses presided over the opposite wall: not the faces of the horses but their rear ends. Next to the painting, Lucite tombstones memorialized mergers and acquisitions, the corporate nuggets Synagro had digested with the help of such investment bankers as Lehman Brothers and Donaldson, Lufkin & Jenrette.

I sat surrounded by gleaming trophies, freshly cut flowers, a crystal vase of potpourri, and a good deal of human waste. Strewn across the table lay Ziplocs of contractor's compost, soil conditioner, seed cover, landscape mulch, All-Gro, and Biogran. On the floor near the podium reclined a huge paper sack of "Hou-Actinite — 100% natural organic fertilizer," with its logo of a golf ball perched on the edge of a flagged hole. My eyes lingered over a jar of pellets labeled NYOFCO. From New York City. *My pellets.*

About this time, Robert Boucher walked into the room — no suit, no tie, clear blue eyes, and a Starbucks of iced green tea. We shook hands, then sat in silence. I had less than an hour and a long list of topics to cover, but before we could get to my questions we needed Alvin Thomas, Synagro's chief legal counsel. Boucher had insisted we have a lawyer present for the interview, and I had some suspicion why: aside from a variety of human-waste-related litigation, Standard & Poor's had recently downgraded Synagro's corporate credit rating on $287 million of outstanding debt. There was plenty to be paranoid about and, as it turned out, a great deal to keep under wraps. A few months after I left Houston, one of the world's largest private equity firms, the Carlyle Group, purchased Synagro Technologies for $776 million.

Robert Boucher grew up in Dover, New Hampshire — "117 miles of sewer lines and one wastewater treatment plant," he said. His father sold garbage trucks, containers, and equipment, and Bob worked summers for Dad, cleaning out used cans and repainting them. He went to Northeastern to play offensive guard but soon dropped out and went to work in the family business.

A company called American Waste gave him his first stab at management, and Boucher began to ascend the corporate ladder. He eventually landed a job at Allied Waste, one of the largest general

waste companies in America. This was big-money garbage, and
Boucher found himself in charge of $1.5 billion in revenues. After
he'd spent a number of years at Allied, bankers came knocking on
his door. They recognized a man they could trust with the future,
and they made Boucher chief operating officer of Synagro. In 2003
he became CEO.

I asked how he perceived human waste in terms of the U.S. economy.

"I think of it from the service aspect," said Boucher. "Running
my business as efficiently as possible keeps our shareholders happy.
Our customers are happy. Our bankers are happy."

But what about the actual substance?

"The material we handle is tougher than garbage," he said. "People don't want to think about it. When you flush your toilet, you
take it for granted — until the day it comes back at you. Then you
have to deal with it. When you go to the commode, that's the end
of the process for you but the beginning for us. We make it go
away."

I told him that Roman and Saxon soothsayers believed they
could prophesy through the art of scatomancie. Like those ancient
adepts, could Boucher analyze the shape of human waste and predict the future?

"We handle material that not a lot of people are interested in today," he said. "It's a blocking-and-tackling-type business, a plug-
along. Very utility-like. It's not sexy."

Clearly, Boucher was evading the question. He did not want to
let on about the car fuel and cement.

"How about energy?" I asked. "When would waste emerge as an
alternative to gasoline?"

He told me that in Europe they burn it, which produces the
same amount of energy as slag coal. Maybe 5,000 British thermal
units per cubic pound. Then he lifted the glass of NYOFCO pellets.
My pellets.

"Let's get our arms around this," he said. "I can't tell you this is
an energy business from the standpoint of reusable fuel. There's
not enough volume."

Not enough volume?

Boucher shook his head.

That took a while to sink in. Not enough volume meant no filling
up the tank. No light bulbs.

"No light bulbs," said Boucher.

But what about big pharma? What about DEET and Chanel?
He shook his head.

No penicillin recovery? No decaffeinated sludge?
He shook his head.

What about metal reclamation? Wasn't this pellet some sort of renewable source of aluminum and chromium?

"It's almost a nondetect," said Boucher.

Clearly, he took me for a fool. He had plenty of reasons to conceal what he and his cronies were up to. He was playing dumb, and I was falling for it. His lawyer looked up from his notes and waited for my next question. What was my next question? I checked the list. What about building materials?

"It turned out not to be so cost-effective," he said. "Land applications have been done forever because they make sense."

I demanded that he prophesy.

"How far into the future?"

"Twenty years," I said.

"Everyone wants the black box to make it disappear," he said. "But that's not what happens. One hundred years from now will they have iPods that make pellets? Sure. Apple will have bought us. They'll have the vapor box." Then his voice dipped beneath sarcasm. "Twenty years is not enough. Twenty years is nothing."

And he gave me the look. Perhaps he was appalled by my ignorance, perhaps he didn't care, but he understood his product and knew the notion of its glorious future was but another symptom of our desire to deny. Despite the primitive and absurd fantasy that we might refine what lay dark within ourselves and reform it into something fabulous, there was no glorious future of waste. There was only this world of shit.

I had no more questions but could not bring myself to leave. So we sat without saying anything for a long time. Eventually, Boucher began to reminisce about first-class travel across America. He flew a lot, and whenever he took his seat on a plane the small talk would begin and the question would arise: what do you do for a living?

"I tell them," recalled Boucher. "And inevitably they say, 'Wow. What a great business.'"

VIRGINIA MORELL

Minds of Their Own

FROM *National Geographic*

IN 1977 Irene Pepperberg, a recent graduate of Harvard University, did something very bold. At a time when animals still were considered automatons, she set out to find what was on another creature's mind by talking to it. She brought a one-year-old African gray parrot she named Alex into her lab to teach him to reproduce the sounds of the English language. "I thought if he learned to communicate, I could ask him questions about how he sees the world."

When Pepperberg began her dialogue with Alex, who died last September at the age of thirty-one, many scientists believed animals were incapable of any thought. They were simply machines, robots programmed to react to stimuli but lacking the ability to think or feel. Any pet owner would disagree. We see the love in our dog's eyes and know that of course Spot has thoughts and emotions. But such claims remain highly controversial. Gut instinct is not science, and it is all too easy to project human thoughts and feelings onto another creature. How, then, does a scientist prove that an animal is capable of thinking — that it is able to acquire information about the world and act on it?

"That's why I started my studies with Alex," Pepperberg said. They were seated — she at her desk, he on top of his cage — in her lab, a windowless room about the size of a boxcar, at Brandeis University. Newspapers lined the floor; baskets of bright toys were stacked on the shelves. They were clearly a team — and because of their work, the notion that animals can think is no longer so fanciful.

Certain skills are considered key signs of higher mental abilities:

good memory, a grasp of grammar and symbols, self-awareness, understanding others' motives, imitating others, and being creative. Bit by bit, in ingenious experiments, researchers have documented these talents in other species, gradually chipping away at what we thought made human beings distinctive while offering a glimpse of where our own abilities came from. Scrub jays know that other jays are thieves and that stashed food can spoil; sheep can recognize faces; chimpanzees use a variety of tools to probe termite mounds and even use weapons to hunt small mammals; dolphins can imitate human postures; the archerfish, which stuns insects with a sudden blast of water, can learn how to aim its squirt simply by watching an experienced fish perform the task. And Alex the parrot turned out to be a surprisingly good talker.

Thirty years after the Alex studies began, Pepperberg and a changing collection of assistants were still giving him English lessons. The humans, along with two younger parrots, also served as Alex's flock, providing the social input all parrots crave. Like any flock, this one — as small as it was — had its share of drama. Alex dominated his fellow parrots, acted huffy at times around Pepperberg, tolerated the other female humans, and fell to pieces over a male assistant who dropped by for a visit. ("If you were a man," Pepperberg said, after noting Alex's aloofness toward me, "he'd be on your shoulder in a second, barfing cashews in your ear.")

Pepperberg bought Alex in a Chicago pet store. She let the store's assistant pick him out because she didn't want other scientists saying later that she'd deliberately chosen an especially smart bird for her work. Given that Alex's brain was the size of a shelled walnut, most researchers thought Pepperberg's interspecies communication study would be futile.

"Some people actually called me crazy for trying this," she said. "Scientists thought that chimpanzees were better subjects, although, of course, chimps can't speak."

Chimpanzees, bonobos, and gorillas have been taught to use sign language and symbols to communicate with us, often with impressive results. The bonobo Kanzi, for instance, carries his symbol-communication board with him so he can "talk" to his human researchers, and he has invented combinations of symbols to express his thoughts. Nevertheless, this is not the same thing as having an animal look up at you, open his mouth, and speak.

Pepperberg walked to the back of the room, where Alex sat on top of his cage preening his pearl gray feathers. He stopped at her approach and opened his beak.

"Want grape," Alex said.

"He hasn't had his breakfast yet," Pepperberg explained, "so he's a little put out."

Alex returned to preening, while an assistant prepared a bowl of grapes, green beans, apple and banana slices, and corn on the cob.

Under Pepperberg's patient tutelage, Alex learned how to use his vocal tract to imitate almost one hundred English words, including the sounds for all of these foods, although he calls an apple a "ban-erry."

"Apples taste a little bit like bananas to him, and they look a little bit like cherries, so Alex made up that word for them," Pepperberg said.

Alex could count to six and was learning the sounds for seven and eight.

"I'm sure he already knows both numbers," Pepperberg said. "He'll probably be able to count to ten, but he's still learning to say the words. It takes far more time to teach him certain sounds than I ever imagined."

After breakfast, Alex preened again, keeping an eye on the flock. Every so often, he leaned forward and opened his beak: "Ssse . . . won."

"That's good, Alex," Pepperberg said. "Seven. The number is seven."

"Ssse . . . won! Se . . . won!"

"He's practicing," she explained. "That's how he learns. He's thinking about how to say that word, how to use his vocal tract to make the correct sound."

It sounded a bit mad, the idea of a bird having lessons to practice and willingly doing it. But after listening to and watching Alex, it was difficult to argue with Pepperberg's explanation for his behaviors. She wasn't handing him treats for the repetitious work or rapping him on the claws to make him say the sounds.

"He has to hear the words over and over before he can correctly imitate them," Pepperberg said, after pronouncing "seven" for Alex a good dozen times in a row. "I'm not trying to see if Alex can learn a human language," she added. "That's never been the point.

My plan always was to use his imitative skills to get a better understanding of avian cognition."

In other words, because Alex was able to produce a close approximation of the sounds of some English words, Pepperberg could ask him questions about a bird's basic understanding of the world. She couldn't ask him what he was thinking about, but she could ask him about his knowledge of numbers, shapes, and colors. To demonstrate, Pepperberg carried Alex on her arm to a tall wooden perch in the middle of the room. She then retrieved a green key and a small green cup from a basket on a shelf. She held up the two items to Alex's eye.

"What's same?" she asked.

Without hesitation, Alex's beak opened: "Co-lor."

"What's different?" Pepperberg asked.

"Shape," Alex said. His voice had the digitized sound of a cartoon character. Since parrots lack lips (another reason it was difficult for Alex to pronounce some sounds, such as *ba*), the words seemed to come from the air around him, as if a ventriloquist were speaking. But the words — and what can only be called the thoughts — were entirely his.

For the next twenty minutes, Alex ran through his tests, distinguishing colors, shapes, sizes, and materials (wool versus wood versus metal). He did some simple arithmetic, such as counting the yellow toy blocks among a pile of mixed hues.

And, then, as if to offer final proof of the mind inside his bird's brain, Alex spoke up. "Talk clearly!" he commanded when one of the younger birds Pepperberg was also teaching mispronounced the word "green." "Talk clearly!"

"Don't be a smart aleck," Pepperberg said, shaking her head at him. "He knows all this, and he gets bored, so he interrupts the others, or he gives the wrong answer just to be obstinate. At this stage, he's like a teenage son; he's moody, and I'm never sure what he'll do."

"Wanna go tree," Alex said in a tiny voice.

Alex had lived his entire life in captivity, but he knew that beyond the lab's door there was a hallway and a tall window framing a leafy elm tree. He liked to see the tree, so Pepperberg put her hand out for him to climb aboard. She walked him down the hall into the tree's green light.

"Good boy! Good birdie," Alex said, bobbing on her hand.

"Yes, you're a good boy. You're a good birdie." And she kissed his feathered head.

He was a good birdie until the end, and Pepperberg was happy to report that when he died he had finally mastered "seven."

Many of Alex's cognitive skills, such as his ability to understand the concepts of same and different, are generally ascribed only to higher mammals, particularly primates. But parrots, like great apes (and humans), live a long time in complex societies. And like primates, these birds must keep track of the dynamics of changing relationships and environments.

"They need to be able to distinguish colors to know when a fruit is ripe or unripe," Pepperberg noted. "They need to categorize things — what's edible, what isn't — and to know the shapes of predators. And it helps to have a concept of numbers if you need to keep track of your flock, and to know who's single and who's paired up. For a long-lived bird, you can't do all of this with instinct; cognition must be involved."

Being able mentally to divide the world into simple abstract categories would seem a valuable skill for many organisms. Is that ability, then, part of the evolutionary drive that led to human intelligence?

Charles Darwin, who attempted to explain how human intelligence developed, extended his theory of evolution to the human brain: like the rest of our physiology, intelligence must have evolved from simpler organisms, since all animals face the same general challenges of life. They need to find mates, food, and a path through the woods, sea, or sky — tasks that Darwin argued require problem-solving and categorizing abilities. Indeed, Darwin went so far as to suggest that earthworms are cognitive beings because, based on his close observations, they have to make judgments about the kinds of leafy matter they use to block their tunnels. He hadn't expected to find thinking invertebrates and remarked that the hint of earthworm intelligence "has surprised me more than anything else in regard to worms."

To Darwin, the earthworm discovery demonstrated that degrees of intelligence could be found throughout the animal kingdom. But the Darwinian approach to animal intelligence was cast aside in the early twentieth century, when researchers decided that field observations were simply "anecdotes," usually tainted by anthropomorphism. In an effort to be more rigorous, many embraced behaviorism, which regarded animals as little more than machines, and focused their studies on the laboratory white rat — since one "machine" would behave like any other.

But if animals are simply machines, how can the appearance of human intelligence be explained? Without Darwin's evolutionary perspective, the greater cognitive skills of people did not make sense biologically. Slowly the pendulum has swung away from the animal-as-machine model and back toward Darwin. A whole range of animal studies now suggests that the roots of cognition are deep, widespread, and highly malleable.

Just how easily new mental skills can evolve is perhaps best illustrated by dogs. Most owners talk to their dogs and expect them to understand. But this canine talent wasn't fully appreciated until a border collie named Rico appeared on a German TV game show in 2001. Rico knew the names of some two hundred toys and acquired the names of new ones with ease.

Researchers at the Max Planck Institute for Evolutionary Anthropology in Leipzig heard about Rico and arranged a meeting with him and his owners. That led to a scientific report revealing Rico's uncanny language ability: he could learn and remember words as quickly as a toddler. Other scientists had shown that two-year-old children — who acquire around ten new words a day — have an innate set of principles that guide this task. The ability is seen as one of the key building blocks in language acquisition. The Max Planck scientists suspect that the same principles guide Rico's word learning and that the technique he uses for learning words is identical to that of humans.

To find more examples, the scientists read all the letters from hundreds of people claiming that their dogs had Rico's talent. In fact, only two — both border collies — had comparable skills. One of them — the researchers call her Betsy — has a vocabulary of more than three hundred words.

"Even our closest relatives, the great apes, can't do what Betsy can do — hear a word only once or twice and know that the acoustic pattern stands for something," said Juliane Kaminski, a cognitive psychologist who worked with Rico and is now studying Betsy. She and her colleague Sebastian Tempelmann had come to Betsy's home in Vienna to give her a fresh battery of tests. Kaminski petted Betsy, while Tempelmann set up a video camera.

"Dogs' understanding of human forms of communication is something new that has evolved," Kaminski said, "something that's developed in them because of their long association with humans." Although Kaminski has not yet tested wolves, she doubts they have this language skill. "Maybe these collies are especially good at it because they're working dogs and highly motivated, and in their traditional herding jobs, they must listen very closely to their owners."

Scientists think that dogs were domesticated about 15,000 years ago, a relatively short time in which to evolve language skills. But how similar are these skills to those of humans? For abstract thinking, we employ symbols, letting one thing stand for another. Kaminski and Tempelmann were testing whether dogs can do this too.

Betsy's owner — whose pseudonym is Schaefer — summoned Betsy, who obediently stretched out at Schaefer's feet, eyes fixed on her face. Whenever Schaefer spoke, Betsy attentively cocked her head from side to side.

Kaminski handed Schaefer a stack of color photographs and asked her to choose one. Each image depicted a dog's toy against a white background — toys Betsy had never seen before. They weren't actual toys; they were only images of toys. Could Betsy connect a two-dimensional picture to a three-dimensional object?

Schaefer held up a picture of a fuzzy, rainbow-colored Frisbee and urged Betsy to find it. Betsy studied the photograph and Schaefer's face, then ran into the kitchen, where the Frisbee was placed among three other toys and photographs of each toy. Betsy brought either the Frisbee or the photograph of the Frisbee to Schaefer every time.

"It wouldn't have been wrong if she'd just brought the photograph," Kaminski said. "But I think Betsy can use a picture, without a name, to find an object. Still, it will take many more tests to prove this."

Even then, Kaminski is unsure that other scientists will ever accept her discovery because Betsy's abstract skill, as minor as it may seem to us, may tread all too closely to human thinking.

Still, we remain the inventive species. No other animal has built skyscrapers, written sonnets, or made a computer. Yet animal researchers say that creativity, like other forms of intelligence, did not simply spring from nothingness. It, too, has evolved.

"People were surprised to discover that chimpanzees make tools," said Alex Kacelnik, a behavioral ecologist at Oxford University, referring to the straws and sticks chimpanzees shape to pull termites from their nests. "But people also thought, 'Well, they share our ancestry — of course they're smart.' Now we're finding these kinds of exceptional behaviors in some species of birds. But we don't have a recently shared ancestry with birds. Their evolutionary history is very different; our last common ancestor with all birds was a reptile that lived over 300 million years ago.

"This is not trivial," Kacelnik continued. "It means that evolution can invent similar forms of advanced intelligence more than once — that it's not something reserved only for primates or mammals."

Kacelnik and his colleagues are studying one of these smart species, the New Caledonian crow, which lives in the forests of that Pacific island. New Caledonian crows are among the most skilled of tool-making and tool-using birds, forming probes and hooks from sticks and leaf stems to poke into the crowns of the palm trees, where fat grubs hide. Since these birds, like chimpanzees, make and use tools, researchers can look for similarities in the evolutionary processes that shaped their brains. Something about the environments of both species favored the evolution of tool-making neural powers.

But is their use of tools rigid and limited, or can they be inventive? Do they have what researchers call mental flexibility? Chimpanzees certainly do. In the wild, a chimpanzee may use four sticks of different sizes to extract the honey from a bee's nest. And in captivity, they can figure out how to position several boxes so they can retrieve a banana hanging from a rope.

Answering that question for New Caledonian crows — extremely shy birds — wasn't easy. Even after years of observing them in the wild, researchers couldn't determine if the birds' ability was innate or if they learned to make and use their tools by watching one another. If it was a genetically inherited skill, could they, like the chimps, use their talent in different, creative ways?

To find out, Kacelnik and his students brought twenty-three

crows of varying ages (all but one caught in the wild) to the aviary in his Oxford lab and let them mate. Four hatchlings were raised in captivity, and all were carefully kept away from the adults, so they had no opportunity to be taught about tools. Yet soon after they fledged, all picked up sticks to probe busily into cracks and shaped different materials into tools. "So we know that at least the bases of tool use are inherited," Kacelnik said. "And now the question is, what else can they do with tools?"

Plenty. In his office, Kacelnik played a video of a test he'd done with one of the wild-caught crows, Betty, who had died recently from an infection. In the film, Betty flies into a room. She's a glossy black bird with a crow's bright, inquisitive eyes, and she immediately spies the test before her: a glass tube with a tiny basket lodged in its center. The basket holds a bit of meat. The scientists had placed two pieces of wire in the room. One was bent into a hook, the other was straight. They figured Betty would choose the hook to lift the basket by its handle.

But experiments don't always go according to plan. Another crow had stolen the hook before Betty could find it. Betty is undeterred. She looks at the meat in the basket, then spots the straight piece of wire. She picks it up with her beak, pushes one end into a crack in the floor, and uses her beak to bend the other end into a hook. Thus armed, she lifts the basket out of the tube.

"This was the first time Betty had ever seen a piece of wire like this," Kacelnik said. "But she knew she could use it to make a hook and exactly where she needed to bend it to make the size she needed."

They gave Betty other tests, each requiring a slightly different solution, such as making a hook out of a flat piece of aluminum rather than a wire. Each time, Betty invented a new tool and solved the problem. "It means she had a mental representation of what it was she wanted to make. Now that," Kacelnik said, "is a major kind of cognitive sophistication."

This is the larger lesson of animal cognition research: it humbles us. We are not alone in our ability to invent or plan or to contemplate ourselves — or even to plot and lie.

Deceptive acts require a complicated form of thinking, since you must be able to attribute intentions to the other person and predict that person's behavior. One school of thought argues that human intelligence evolved partly because of the pressure of living in a complex society of calculating beings. Chimpanzees, orangutans, gorillas, and bonobos share this capacity with us. In the wild, primatologists have seen apes hide food from the alpha male or have sex behind his back.

Birds, too, can cheat. Laboratory studies show that western scrub jays can know another bird's intentions and act on that knowledge. A jay that has stolen food itself, for example, knows that if another jay watches it hide a nut, there's a chance the nut will be stolen. So the first jay will return to move the nut when the other jay is gone.

"It's some of the best evidence so far of experience projection in another species," said Nicky Clayton in her aviary lab at Cambridge University. "I would describe it as 'I know that you know where I have hidden my stash of food, and if I were in your shoes I'd steal it, so I'm going to move my stash to a place you don't know about.'"

This study, by Clayton and her colleague Nathan Emery, is the first to show the kinds of ecological pressures, such as the need to hide food for winter use, that would lead to the evolution of such mental abilities. Most provocatively, her research demonstrates that some birds possess what is often considered another uniquely human skill: the ability to recall a specific past event. Scrub jays, for example, seem to know how long ago they cached a particular kind of food, and they manage to retrieve it before it spoils.

Human cognitive psychologists call this kind of memory "episodic memory" and argue that it can exist only in a species that can mentally travel back in time. Despite Clayton's studies, some refuse to concede this ability to the jays. "Animals are stuck in time," explained Sara Shettleworth, a comparative psychologist at the University of Toronto in Canada, meaning that they don't distinguish among past, present, and future the way humans do. Since animals lack language, she said, they probably also lack "the extra layer of imagination and explanation" that provides the running mental narrative accompanying our actions.

Such skepticism is a challenge for Clayton. "We have good evidence that the jays remember the what, where, and when of specific caching events, which is the original definition of episodic memory. But now the goalposts have moved." It's a common complaint among animal researchers. Whenever they find that a species has a mental skill reminiscent of a special human ability, the

human-cognition scientists change the definition. But the animal
researchers may underestimate their power — it is their discoveries
that compel the human side to shore up the divide.

"Sometimes the human-cognitive psychologists can be so fixed
on their definitions that they forget how fabulous these animal discoveries are," said Clive Wynne of the University of Florida, who
has studied cognition in pigeons and marsupials. "We're glimpsing
intelligence throughout the animal kingdom, which is what we
should expect. It's a bush, not a single-trunk tree with a line leading only to us."

Some of the branches on that bush have led to such degrees of
intelligence that we should blush for ever having thought any animal a mere machine.

In the late 1960s a cognitive psychologist named Louis Herman
began investigating the cognitive abilities of bottlenose dolphins.
Like humans, dolphins are highly social and cosmopolitan, living
in subpolar to tropical environments worldwide; they're highly vocal; and they have special sensory skills, such as echolocation. By
the 1980s Herman's cognitive studies were focused on a group of
four young dolphins — Akeakamai, Phoenix, Elele, and Hiapo —
at the Kewalo Basin Marine Mammal Laboratory in Hawaii. The
dolphins were curious and playful, and they transferred their socia-
bility to Herman and his students.

"In our work with the dolphins, we had a guiding philosophy,"
Herman says, "that we could bring out the full flower of their intellect, just as educators try to bring out the full potential of a human
child. Dolphins have these big, highly complex brains. My thought
was, 'OK, so you have this pretty brain. Let's see what you can do
with it.'"

To communicate with the dolphins, Herman and his team invented a hand- and arm-signal language, complete with a simple
grammar. For instance, a pumping motion of the closed fists meant
"hoop," and both arms extended overhead (as in jumping jacks)
meant "ball." A "come here" gesture with a single arm told them to
"fetch." Responding to the request "hoop, ball, fetch," Akeakamai
would push the ball to the hoop. But if the word order was changed
to "ball, hoop, fetch," she would carry the hoop to the ball. Over
time she could interpret more grammatically complex requests,

such as "right, basket, left, Frisbee, in," asking that she put the Frisbee on her left in the basket on her right. Reversing "left" and "right" in the instruction would reverse Akeakamai's actions. Akeakamai could complete such requests the first time they were made, showing a deep understanding of the grammar of the language.

"They're a very vocal species," Herman adds. "Our studies showed that they could imitate arbitrary sounds that we broadcast into their tank, an ability that may be tied to their own need to communicate. I'm not saying they have a dolphin language. But they are capable of understanding the novel instructions that we convey to them in a tutored language; their brains have that ability.

"There are many things they could do that people have always doubted about animals. For example, they correctly interpreted, on the very first occasion, gestured instructions given by a person displayed on a TV screen behind an underwater window. They recognized that television images were representations of the real world that could be acted on in the same way as in the real world."

They readily imitated motor behaviors of their instructors too. If a trainer bent backward and lifted a leg, the dolphin would turn on its back and lift its tail in the air. Although imitation was once regarded as a simple-minded skill, in recent years cognitive scientists have revealed that it's extremely difficult, requiring the imitator to form a mental image of the other person's body and pose, then adjust his own body parts into the same position — actions that imply an awareness of one's self.

"Here's Elele," Herman says, showing a film of her following a trainer's directions. "Surfboard, dorsal fin, touch." Instantly Elele swam to the board and, leaning to one side, gently laid her dorsal fin on it, an untrained behavior. The trainer stretched her arms straight up, signaling "Hooray!" and Elele leaped into the air, squeaking and clicking with delight.

"Elele just loved to be right," Herman said. "And she loved inventing things. We made up a sign for 'create,' which asked a dolphin to create its own behavior."

Dolphins often synchronize their movements in the wild, such as leaping and diving side by side, but scientists don't know what signal they use to stay so tightly coordinated. Herman thought he might be able to tease out the technique with his pupils. In the film, Akeakamai and Phoenix are asked to create a trick and do it together. The two dolphins swim away from the side of the pool, circle together underwater for about ten seconds, then leap out of the water, spinning clockwise on their long axis and squirting water from their mouths, every maneuver done at the same instant. "None of this was trained," Herman says, "and it looks to us absolutely mysterious. We don't know how they do it — or did it."

He never will. Akeakamai and Phoenix and the two others died accidentally four years ago. Through these dolphins, he made some of the most extraordinary breakthroughs ever in understanding another species' mind — a species that even Herman describes as "alien," given its aquatic life and the fact that dolphins and primates diverged millions of years ago. "That kind of cognitive convergence suggests there must be some similar pressures selecting for intellect," Herman said. "We don't share their biology or ecology. That leaves social similarities — the need to establish relationships and alliances superimposed on a lengthy period of maternal care and longevity — as the likely common driving force."

"I loved our dolphins," Herman says, "as I'm sure you love your pets. But it was more than that, more than the love you have for a pet. The dolphins were our colleagues. That's the only word that fits. They were our partners in this research, guiding us into all the capabilities of their minds. When they died, it was like losing our children."

Herman pulled a photograph from his file. In it he is in the pool with Phoenix, who rests her head on his shoulder. He is smiling and reaching back to embrace her. She is sleek and silvery with appealingly large eyes, and she looks to be smiling too, as dolphins always do. It's an image of love between two beings. In that pool, at least for that moment, there was clearly a meeting of the minds.

J. MADELEINE NASH

Back to the Future

FROM *High Country News*

CLAMBERING ONTO A DUN-COLORED KNOLL, not far from the small town of Worland, Wyoming, Scott Wing stares out at the deeply abraded hills that sweep toward him like the waves of a vast stony ocean. "That's it," he says, pointing to a sinuous ribbon of rose-colored rock. "That's the Big Red." I follow his gaze, noting how the Big Red snakes into an arroyo, then disappears around a bend. Even to my untrained eye, the geological band seems to glow with a fierce, otherworldly intensity.

In some places, Wing explains, the Big Red is composed of multiple stripes; in others, it wends through the landscape as a single line of color. Then, too, the capricious hand of erosion has exposed it here, left it hidden over there. But after hours of pondering the pieces of this jigsaw, Wing, a paleobotanist at the Smithsonian's Museum of Natural History in Washington, D.C., believes he can now follow the Big Red for a distance of twenty-five miles, from the base of the solitary outcrop that looms in the distance all the way to the Sand Creek Divide.

The Sand Creek Divide is a high point in Wyoming's Big Horn Basin. From it you can see the emerald patchwork of irrigated sugar beet and malt barley fields that hug the Big Horn River as well as the jagged mountain ranges — the Absarokas, the Big Horns, the Owl Creeks — that define the edges of this harsh mid-latitude desert. Temperatures here regularly dip below 0 degrees Fahrenheit in wintertime and, in summer, soar well past 100. Away from waterways, the vegetation amounts to little more than a stippling of sagebrush intermingled with stands of invasive cheatgrass and ephemerally blooming wildflowers.

But between 55 and 56 million years ago, says Wing, the Big Horn Basin was a balmy, swampy Eden, teeming with flora and fauna that would be at home in today's coastal Carolinas. Crocodiles, turtles, and alligator gar plied the waters of meandering rivers, and early mammals scampered through woodlands filled with the relatives of modern sycamores, bald cypresses, and palms. And then, all of a sudden, things got a whole lot warmer. In a geological eye blink — less than 10,000 years, some think — global mean temperatures shot up by around 10 degrees Fahrenheit, jump-starting a planetary heat wave that lasted for over 150,000 years.

Here, in the southeastern sector of the basin, the Big Red is the most vivid marker of this exceptionally torrid time — the Paleocene-Eocene Thermal Maximum, or PETM, as most paleontologists call it. By following the Big Red, Wing and his colleagues hope to locate fossils and other clues that will help them reconstruct this long-vanished world — a world with unexpected relevance for us as we hurtle toward our own rendezvous with climate change.

Scientists believe that then, as now, Earth warmed in response to a precipitous release of carbon dioxide and other heat-trapping gases, setting in motion events that reverberated through both marine and terrestrial ecosystems. But where did those gases come from so long ago? What triggered their sudden release? And, most important of all, how likely is it that the PETM, or something disquietingly close to it, could happen all over again?

In 1972, when the seventeen-year-old Wing made the first of many trips to the Big Horn Basin, scientists knew too little even to frame such questions. Today, however, dozens of paleontologists, oceanographers, geochemists, and climate modelers are racing to come up with answers. Nowhere have they struck a more productive lode than in these candy-striped badlands. As Wing says, "You can literally walk up to a layer of rock and know that the Paleocene-Eocene boundary starts *here.*"

Leaning on a long-handled shovel, Wing goes over the field schedule with a couple of colleagues, then heads back to Dino, a rust-colored 1970 Suburban with a birdlike dinosaur painted on each side. Wing bought this unlikely chariot in 1987 and somehow has kept it running ever since.

Five minutes later, he pulls up to the site that everyone refers to as "Ross's quarry" in honor of the University of Nebraska paleontologist Ross Secord, who discovered it last year. Wing's crew has formed a conga line of shovelers, and as their fifty-three-year-old leader scrambles up from below, they fling clouds of grit in his direction. Eventually, the pace of shoveling slows down so that promising chunks of rock can be individually examined and, if necessary, split open with a hammer. The best specimens are passed to Wing, who peers at each one through his eyepiece and decides whether to keep or discard it.

"This is a good one," Wing calls, so I climb up to see. On the surface of the rock is an exquisitely formed leaf, its veins and margins perfectly preserved. Grayish brown in color and slightly dank, the 55-million-year-old leaf looks as if it might have fallen last week and is just now beginning to molder. Adding to the illusion of freshness, its fossilized tissue retains traces of the waxes that once comprised its protective exterior coating.

The plant to which this leaf once belonged, Wing thinks, migrated from far to the south in response to warming temperatures. Like a time capsule, the leaf carries information that can illuminate what it was like to live in a rapidly warming world.

"So far, what we've learned is that processes we're now affecting are so complicated that we can't easily model them," Wing says. "We can monitor them, but over short periods of time there's so much noise in the system that it overwhelms the signal. That's why the geological and paleontological record is so important. It's one of the few ways we can look into how the system works." With that, Wing turns away to squint at another leaf. Unshaven, with a broad-brimmed hat squashed onto his head and a notebook stuffed into a field vest pocket, he looks just like the seasoned fossil hunter he is.

Even before it had a name, the Paleocene-Eocene Thermal Maximum was starting to fascinate Wing. For some time it had been clear to paleontologists studying the evolution of mammals that the transition between the Paleocene and the Eocene was marked by the kind of innovative burst that implies sweeping ecological change. Yet no hint of such a change had appeared in any of the fossil leaves Wing had collected. He would stare at leaves from the Paleocene and leaves from the Eocene but see almost no difference between them. "It was getting to be annoying," he recalls.

The Paleocene is the geological epoch that started 65 million years ago, right after a wayward asteroid or comet crashed into the planet, ending the reign of the dinosaurs. At the time, mammals were rather simple, general-purpose creatures with few specializations: their teeth, ankle bones, and joints all look extremely primitive. Then, barely 10 million years later, at the dawn of the Eocene, the first relatives of deer abruptly appear, along with the first primates and first horses.

"You can literally draw a line through the rock," says Philip Gingerich, a vertebrate paleontologist at the University of Michigan. "Above it there are horses; below it there aren't." In fact, where Gingerich works — at Polecat Bench, in the northern sector of the Big Horn Basin — you can actually see the line, in the form of a band of light gray sandstone. Oddly enough, many fossil mammals commonly found above this line, including those first horses, were abnormally small. Typically, Gingerich says, Eocene horses grew to the size of modern-day cocker spaniels, but these horses were "about the size of Siamese cats."

In 1991, as Gingerich and others were marveling over the miniature mammals of Polecat Bench, the oceanographers James Kennett of the University of California, Santa Barbara, and Lowell Stott of the University of Southern California investigated a major extinction of small, shelly creatures that during the late Paleocene lived on the sea floor off the coast of Antarctica. This massive die-off, they found, coincided with a steep rise in deep-ocean temperatures and a curious spike in atmospheric carbon.

Less than a year later, the paleontologist Paul Koch and the paleo-oceanographer James Zachos, both now at the University of California, Santa Cruz, teamed up with Gingerich to show that this geochemical glitch had also left its calling card on land. The trio established this indirectly by measuring the carbon content of fossilized teeth and nodules plucked from the Big Horn Basin's 55.5-million-year-old rocks.

To Wing, it began to seem increasingly implausible that plant communities could have segued through the PETM unaffected. So in 1994 he started a methodical search for the fossils that he was all but sure he had missed, returning year after year to the Big Horn Basin. He started in its southeastern corner and then moved north to explore Polecat Bench and the Clarks Fork Basin. Yet it wasn't until 2003, when he reached the Worland area, that he began to meet with success.

At first he found just a smattering of leaves, too few to suggest any pattern. Then in 2005, at the end of a long day, he slid his shovel into a grayish mound and pulled out a tiny leaf. "I knew immediately that this was totally different from anything I'd seen before, that this was really dramatic, so I got down on my knees and poked the shovel in again and then again. In every shovelful, there were more leaves coming out. First I started to laugh; then I started to cry. And then I looked up."

Staring down at Wing was a new field assistant. "He had a look on his face that said, 'Now I'm going to die.' I understood what he must have been thinking. 'Here I am, I've just graduated from college. I've never camped before. I've never been on a paleontological expedition before. It's six o'clock in the evening. It's still one hundred degrees. I don't know where I am. And it looks like the boss has gone completely nuts!' So I said to him, 'Really, it's OK. I'm not crazy. It's just that I've been looking for this since you were ten years old!'"

From that one site, Wing went on to extract more than two thousand leaf fossils representing thirty different species. Missing from the mix are the cypresses and other conifers that were so common during the Paleocene; gone also are the distant cousins of broadleaf temperate-zone trees like sycamores, dogwoods, birches. In their place are the legumes, a family of plants, shrubs, and trees — think of acacias and mimosas — that thrive today in seasonally dry tropical and subtropical areas.

"What you see is almost a complete changeover from what was growing here before," Wing marvels. "What this means is that you could have stood in this one spot in Wyoming, surrounded by a forest, and everything would have looked pretty much the same for millions of years. And then, over a few tens of thousands of years, almost all the plants you're familiar with disappear and are replaced by plants you've never seen before in your life." At least some of the newcomers migrated north from as far away as the Mississippi Embayment, precursor of the Gulf of Mexico. With them came a wave of small but voracious predators: many of the fossil leaves are peppered with the scars left by chewing, sucking, mining, and boring insects.

It's a cool, clear morning, with just a few wisps of cirrus streaking the sky, when Francesca Smith settles into a spot just above Ross's quarry, sitting cross-legged on the crackled ground. Across the road, along the ridgeline, are three petroleum pump jacks, their heads slowly rising and dipping as underground reservoirs empty, then fill with oil.

An associate professor at Northwestern University, the bubbly, brown-haired geochemist arrived only yesterday, having driven herself and two young assistants out from Illinois. Wing hands her a slab of mudstone. "Look," he says, pointing to the merest fragment of a leaf splayed across the surface. "That's a little piece of organic matter — it's probably part of the cuticle." At that moment, the wind picks up, and the paper-thin specimen peels back from the rock, threatening to fly away. "Emergency wrap!" Smith shouts, hastily enfolding her prize in an envelope of foam.

For a moment, Smith contemplates the pump jacks, visual metaphors linking the prehistoric world she and Wing are exploring to our world's present and future. The carbon released during the PETM, she notes, is thought to have come from organic sources, just like the carbon we pump into the atmosphere every time we turn on a light or drive a car. "The only difference," Smith reflects, "is that we're doing it much, much faster."

During the Paleocene-Eocene Thermal Maximum, scientists estimate that a massive amount of carbon — 4 to 5 trillion metric tons, perhaps — flooded into the atmosphere. That's about ten times more carbon than humans have pumped out since 1751 and the rough equivalent of how much carbon remains stored in fossil fuels.

From a climatological perspective, it makes sense that the infusion of that much carbon would jack up temperatures. After all, carbon combines with oxygen to form carbon dioxide, which, next to water vapor, is the most abundant of the planet's greenhouse gases. As their name suggests, these gases (which also include methane and nitrous oxide) behave rather like the transparent panes of a greenhouse: they allow the sun's rays to stream in but trap a good deal of the heat that Earth beams back in response.

In general, this is a good thing; it helps create what some call the Goldilocks effect — the fact that to human beings and other creatures, Earth's temperature seems "just right," neither hellishly hot like Venus nor bitterly cold like Mars. That's not to say that our planet's pane of greenhouse gases never varies in thickness. Ancient air bubbles trapped in Antarctica's ice show that levels of carbon dioxide declined during past ice ages and rose during warm interglacials such as our own.

It's not clear how much carbon dioxide there was in the atmosphere on the eve of the PETM, but scientists think levels may have reached somewhere between 500 and 750 parts per million. This compares to 380 parts per million at present, 280 parts per million in preindustrial times, and 180 during past glacial high stands. As a result, the late Paleocene was already quite warm, about as warm as many climatologists project our world could become by the start of the next century.

The large amount of carbon dioxide in the pre-PETM atmosphere almost certainly came from a sustained spate of volcanic eruptions. (During the Paleocene, volcanoes were particularly active.) As a result, the atmosphere's load of carbon dioxide gradually rose. Then, around 55.5 million years ago, carbon dioxide levels shot up very sharply, perhaps to 1,800 or more parts per million, leaving behind a distinctive geochemical signature.

The signature, Smith explains, takes the form of a dramatic shift in the ratio of two stable forms of carbon: heavier carbon 13 and lighter carbon 12. It's this shift that scientists first picked up in the calcareous shells of marine organisms, then found in the teeth of terrestrial mammals. Last year Smith and Wing showed that in the Big Horn Basin the shift is captured by leaf waxes as well. "And there is only one way we know of to shift the ratio as much as it shifted," Smith says, "and that's to add a lot more light carbon."

The richest concentrations of light carbon are found in organic materials, including fossil fuels like coal (which forms from deeply buried plants) and methane gas (primarily a byproduct of microbial decomposition). While scientists are still not sure what triggered the massive release of light carbon at the start of the PETM, they do have a number of possible culprits, including fires that raged through forests, dried-up peat bogs, and even underground coal seams, and effusively erupting volcanoes whose magma intruded into organic-rich sediments, cooking out the carbon.

Of all the scenarios so far floated, perhaps the most provocative invokes the dissolution of methane hydrates on the sea floor. These are icelike solids in which water molecules form crystalline cages that entrap molecules of gas; they form and remain stable within specific ranges of temperature and pressure. At the end of the

Paleocene, the Rice University earth scientist Gerald Dickens has suggested, a jolt of warmth from an unknown source pushed these strange solids to the point that the methane gas inside them started burbling up through the ocean and into the atmosphere.

Methane is much shorter-lived than carbon dioxide, but it's also a more effective greenhouse gas. And as methane breaks down, the carbon it contains recombines with oxygen to form carbon dioxide, which can circulate through the climate system for thousands of years.

After serious study, many experts have concluded that not enough methane was locked up in hydrate form to have single-handedly caused the PETM. That does not constitute an absolution, however. A big release of sea-floor methane could still have been part of a sustained chain reaction whereby an initial rise in carbon caused enough warming to trigger the release of additional carbon that caused still more warming, and so on. Might the carbon we are so heedlessly pumping out today spark a similar sequence of events?

As scientists try to imagine the consequences of our greenhouse gas emissions, they invariably return to this question. The earth abounds with "traps" for carbon — not just seafloor hydrates but also terrestrial forests, marine plankton, and frozen Arctic soils. Some of these our own warming climate may already be springing open. Scientists from the University of Alaska, Fairbanks, recently calculated that the permafrost of the far North sequesters 100 billion tons of carbon in its top three feet alone; as the permafrost thaws, that carbon will progressively leak into the atmosphere.

The release of carbon from some of these traps will be offset by the uptake of carbon in others. But at some point, scientists worry, the release of carbon may so far outweigh its absorption that the situation will cascade out of control. By the time we realize we're in serious trouble, in other words, it may be too late to do much about it.

Inside a cavernous tent bathed in golden late-afternoon light, the resident team of vertebrate paleontologists pores over the day's haul, emitting sporadic whoops of surprise. "That might be a eureka," exclaims Yale University graduate student Stephen Chester, peering through a microscope at a tooth embedded in a jawbone fragment. "That might be a primate."

"It is! It's totally *Teilhardina!*" Doug Boyer, a Ph.D. candidate at New York's Stony Brook University, enthusiastically agrees. *Teilhardina*, he explains, is the Latin name for a group of primates that appeared in Asia, Europe, and North America at roughly the same time.

"That's a horse, Dougie," Chester says, examining another tooth. It belongs to *Hyracotherium sandrae*, the unusually small species first identified at Polecat Bench. The tooth is a shiny dark amber and it's very tiny. Why was this horse, *H. sandrae,* so small?

The most straightforward explanation is that *H. sandrae* was simply a small-size species that migrated into the Big Horn Basin from somewhere else. But the University of Michigan's Gingerich champions a more intriguing possibility. He suggests that its diminutive stature could be the consequence of a decline in available nutrients, notably protein. Horticultural experiments have shown that some plants, when bathed in high concentrations of carbon dioxide, have less protein in their leaves.

The same phenomenon may also have caused insects to become more voracious, which is consistent with the leaf damage displayed by Wing's fossils. Here again, however, there are other possible explanations. For example, higher temperatures alone would have raised insect food requirements by quickening metabolic rates and encouraging year-round breeding. "During the PETM, many things are changing all at once, and it's hard to separate one from another," Wing observes.

At present, Wing is in the field camp's cooking tent, heating up his favorite utensil, a big black wok that can handle dinner for sixteen. Just behind him, seated at a long metal table, Mary Kraus, a wiry sedimentologist from the University of Colorado, is starting to peel a big pile of russet potatoes. She and her daughter, Christina, have had a great day, she says, digging a trench through the pastel paleosols of the badlands surrounding a perennially dry fork of Nowater Creek. "Look at this treasure," she exclaims, holding up a fossilized insect burrow shaped vaguely like a cowboy's boot.

Like tree rings and deep-sea sediments, burrows are what scientists refer to as "proxies." Simply put, proxies are natural systems that record and preserve information about past climates, not unlike modern instruments. Crayfish burrows indicate soils that experience large fluctuations in wetness; earthworm and beetle burrows suggest drier conditions. The colors of ancient soils are also proxies. For example, the degree of redness — whether the color tends toward orange or toward purple — can be correlated with specific ranges of soil moisture.

There are many other types of proxies, including fossil leaves, teeth, and the shells of marine organisms. Typically these proxies record shifts in the ratio between heavy and light elements. Oxygen shifts can be translated into temperature; hydrogen shifts into relative humidity. Changes in leaf size and shape can likewise be read as proxies for temperature and moisture, though as Wing admits, "We don't fully understand why."

Multiple proxies, Wing says, suggest that during the PETM this area of Wyoming was rather similar to South Florida, with a mean annual temperature of around 75 degrees F and annual precipitation between 30 and 55 inches. The precipitation may have followed a strongly seasonal pattern, especially toward the beginning, with part of the year being quite dry, Wing believes. But that's just the broad-brush picture. Wing, Smith, and Kraus, along with the University of Florida paleontologist Jonathan Bloch, who heads up the vertebrate-fossil collection effort, are working on reconstructing the regional climate in much finer detail.

Kraus is using ancient soils to begin mapping what seems like a climatological progression. The initial phase of the PETM looks rather dry, she says, and the middle phase appears drier still, though there are signs of very rapid soil deposition from flooding along rivers and streams. Toward the end of the PETM, in the Big Red itself, she is finding hints that conditions may have become wetter. Among other things, the rocks of the Big Red contain a lot of purple, a color suggestive of higher water tables and more poorly drained soils.

The Big Horn Basin, Kraus says, is probably the ideal place to try to pull together a comprehensive picture of how climate changed on a regional scale over the course of the PETM. "Where else can you go and find 5,000-year intervals stacked one on top of another?" she asks. "Where else can you go and know that 40 meters [about 130 feet] of rock equals 150,000 years?" That's about how long it took the PETM to wind up and wind down, so it's not surprising that over the course of so many millennia, both regional and global climate patterns underwent successive changes. The wind-up, of course, was the fast part; it was the wind-down that took a long time.

Delicious aromas arise from the wok as Wing adds onions and garlic and ginger. Following their noses, the vertebrate fossil crew streams in. Soon we are all sitting in camp chairs, chowing down on rice and spicy curry. After dinner, when the dishes are all washed, dried, and put away, Wing pulls out the battered acoustic guitar he bought for eight dollars years ago in a Worland pawn shop. He starts strumming it softly. Two more members of the group join in, one on a battery-powered keyboard, the other on a tinny guitar.

As the trio warbles out a medley of familiar songs, I contemplate the gossamer sash of the Milky Way as it flows across the nighttime sky. The universe is almost 14 billion years old. Some of the stars in our galaxy are 10 billion years old. The Earth and the sun it circles are around 4.5 billion years old. Measured against such a long stretch of time, the duration of the Paleocene-Eocene Thermal Maximum seems absurdly insignificant. Compared to our own allotment of some few score years and ten, however, it looms a great deal larger. In little more than 150,000 years, ice ages came and went and started anew. In 150,000 years, modern humans diverged from their archaic ancestors and began to spread across the world.

For most of that time, our forebears lived in small, mobile clusters of hunter-gatherers. They began coalescing into settled agricultural communities perhaps 10,000 years ago. Their history as members of a technologically advanced industrial civilization is breathtakingly recent, powered into existence by the eighteenth-century invention of efficient coal-fired steam engines. Who would have predicted that over the span of so few centuries, the clever, adaptable descendants of Eocene primates would become so numerous — and so dependent on carbon-based fuels — as to unbalance the planet?

Perhaps, many hundreds of thousands of years from now, paleontologists from some advanced civilization will uncover fossils from our world and marvel at the carbon shift recorded by the teeth of free-ranging cattle and sheep and the leaves of garden shrubs and trees. Will those beings fathom the real wonder of our story, the fact that we had glimmers of what the future held and yet failed to use that knowledge? Or will our story, like one of Shakespeare's dark comedies, work its way to a happier ending?

It's not that the PETM offers a precise road map to our future, Wing says, when I ask for his thoughts. "It's more that it's an example of the surprises that are waiting for us out there. How was it that Mark Twain put it? History does not repeat itself, but it sure does rhyme."

MICHELLE NIJHUIS

To Take Wildness in Hand

FROM *Orion*

TORREYA STATE PARK perches on the steep, sandy banks of the
Apalachicola, where the river twists slowly through the Florida Panhandle toward the Gulf of Mexico. This is one of the most isolated
spots in Florida, rich only in plant life and prisons, stupefyingly hot
in summer and eerily quiet nearly all year round. Most park visitors
are on their way to somewhere else, and when Connie Barlow
stopped here on a winter day in 1999, she was no exception.

Barlow, trim and now in her fifties, is a writer and naturalist with
cropped hair and a childlike air of enthusiasm. She's given to wandering, and back then she shuttled between a trailer in southern
New Mexico and an apartment in New York City. That winter, during a detour to Florida, she paused at the park for a look at its rai-
son d'être — an ancient tree species called *Torreya taxifolia*, famil-
iarly known as the Florida torreya or, less romantically, stinking
cedar. The park lies at the heart of the tree's tiny range, which
stretches little more than twenty miles from the Georgia state line
toward the mouth of the Apalachicola. But even at Torreya State
Park, Barlow discovered, the Florida torreya is hard to find.

Torreya taxifolia was once a common sight along the Apalachicola, plentiful enough to be cut for Christmas trees, its rot-resistant
wood perfect for fence posts. But at some point in the middle of
the last century — no one is quite sure when — the trees began to
die. Beset by a mysterious disease, overabundant deer, feral hogs,
drought, and perhaps a stressful climate, the adult trees were reduced to a handful of mossy trunks, rotting in riverside ravines.

The species persists in Florida as less than a thousand gangly survivors, most only a few feet tall, their trunks no thicker than a child's wrist, none known to reproduce. Much like the American chestnut, these trees are frozen in preadolescence, knocked back by disease or other adversaries before they grow large enough to set seed. To see their grape-sized seeds, Barlow had to visit the state park offices, where two sit preserved in a jam jar.

Barlow continued her travels that winter but returned to the park a few years later. She tracked down some of the few remaining trees and, in a quiet moment, sat under one of the largest specimens, perhaps ten feet tall. The Florida torreya, even at its healthiest, isn't an obviously charismatic tree. Its flat needles are scanty; its trunk lacks the grandeur of a redwood or an old-growth fir; when it does manage to produce seeds, the rotting results smell like vomit. In its diminished state, it inspires more pity than awe; to call its spindly limbs a canopy is a sorry joke.

But when Barlow looked up at the branches of the Florida torreya, she made an impulsive commitment to the species. She'd spent years thinking and writing about evolution and ecology, and was aware of the implications of climate change. She decided the species needed to move north, to cooler, less diseased climes. And since it couldn't move fast enough alone, Barlow would move it herself.

Climate change is beginning to make good on its threats, and news of its work is now hard to avoid. Escalating average global temperatures? Check. Rising seas? Check. Plants and animals scampering uphill and toward the poles? Check. Dozens of bird and butterfly species are shifting their ranges to cooler terrain or migrating earlier in the year, each species reacting somewhat differently. Ecological communities, never as stable as we might like to think, are disarticulating in new ways.

Conservationists, in response, have offered more ambitious versions of familiar strategies. Bigger nature reserves. More protected corridors for wildlife migration and movement. More regulations, incentives, and ingenuity in service of greenhouse-gas reductions. But even the most expedient tactics could leave some species — especially those as tightly circumscribed as the Florida torreya — marooned in habitat too hot, dry, wet, or stormy.

What then? Captive breeding without hope of reintroduction is an expensive and indefinite custodial project, an ark with no gang-plank. The next option sounds either laughable or desperate: pick up the plants and animals and carry them to better habitat.

Jason McLachlan, an ecologist at the University of Notre Dame, remembers giving a talk in North Carolina about forests' responses to climate change. "Someone in the audience said, 'Why is this a problem? You can just move them,'" he says. "I thought he was just being a smart-ass."

It's an easy idea to caricature. FedEx the polar bears to Antarctica! Airlift the pikas and the orchids! But some scientists take the concept very seriously. Camille Parmesan, a professor at the University of Texas at Austin and an authority on the ecological effects of climate change, remembers broaching the subject at an international conservation conference nearly a decade ago. "I said, 'Look, we need to start thinking about transplanting organisms around these barriers of agricultural land or urban land, and getting them to the next possible suitable habitat as the climate changes,' and people were horrified — just horrified," she says. "They said, 'You can't *do* that!'"

But discussion continued among scientists — if mostly in whispers — and in 2004, a graduate student named Brian Keel quietly coined a term for the idea: assisted migration. Not long afterward, Connie Barlow and the Florida torreya shoved the debate into the open.

The Apalachicola River is bordered by a thick layer of sand, in places more than one hundred feet deep, left when the sea retreated some 2 million years ago. Rain — which fell generously here, exceeding sixty inches each year until the recent drought — hits the loose, sandy soil and keeps going, seeping downward until layers of clay and limestone stop its vertical progress. The moisture then turns toward the main stem of the river, each trickle pulling a few grains of sand with it, a sabotage from below known as sapping erosion.

Over millennia, sapping erosion has created nearly sheer-walled ravines known as steepheads, their sandy banks held in place by magnolias, pines, and muscular beeches. Found in only a handful of other spots throughout the world, steepheads and the shady forests they cradle now define this stretch of the Apalachicola. To step from the sunny, logging-scarred Apalachicola uplands into a steep-head is to enter a darker, wetter, more complicated world, ignored by chainsaws and seemingly hidden from time.

On a humid fall day near the end of hurricane season, David Printiss leads the way over the edge of a steephead, pointing out the faint, narrow path that hairpins down the wall. A few moments after beginning the descent, he crouches in the leaf litter, then turns with a grin. "Introducing *Torreya taxifolia!*"

The tree is a bundle of pencil-thin stems, the tallest two feet high, ridiculously small in comparison to the mature trees surrounding it, dwarfed even by a single leaf of a nearby needle palm.

Printiss is the manager of this preserve — the Nature Conservancy Apalachicola Bluffs and Ravines Preserve, just south of Torreya State Park — and he's proud that the Florida torreya survives in these ravines, even in this almost symbolic state. But he spends most of his time thinking not about the fewer than one hundred *Torreya taxifolia* on the preserve but about the landscape surrounding them. Restoring that, he says, is the best way to solve the "*Torreya* puzzle" and give the tree a chance to thrive.

Printiss has a salt-and-pepper beard, a discreet earring, and a serious demeanor, and he lives here on the preserve with his wife and young daughter. He wears Carhartt work trousers and heavy leather fire boots to the office, and uses both. Each year he serves as "burn boss" on about twenty prescribed fires, some as large as five hundred acres.

"If I can get fire across the landscape acting in its natural role, I've done my job; I'm home," says Printiss as he drives the soft, sandy roads on the flat preserve uplands. "I'm not saying fire is the answer, but I suspect it's a large part of the answer." Restoring fire to the uplands, he says, thins out the overgrown hardwood trees, makes room for the restoration of longleaf pine stands and native grasslands, and brings some filtered sunlight back to the steep ravines where the Florida torreya once grew.

Printiss acknowledges that even if the species were to be revived by these efforts, it could still face the perplexing blight, which attacks the trees by killing the stems and leaving the trees to resprout from their bases. Most surviving *Torreya taxifolia*, like the one at Printiss's feet, have withstood multiple onslaughts and are now clusters of genetically identical stems; since the 1960s, only a single tree is known to have set seed. Despite years of study, no researcher has conclusively identified the disease or its source, and some speculate it may even be a suite of diseases.

Since *T. taxifolia* has separate male and female plants, any trees that managed to persist through adolescence would need the added good fortune of growing near a mate. Only then could the pollen ride the wind to a female tree and produce the species' distinctively hefty seeds. On top of those difficulties is the Southeast's record-breaking drought, which shrank water supplies to dangerously low levels last fall, making the oncoming stresses of climate change difficult to ignore.

Yet the suggestion of assisted migration, of planting *Torreya taxifolia* trees outside these Panhandle steepheads, makes Printiss's face tighten. Such efforts, he says, threaten to take attention and funding away from the work in the preserve and make an already bad situation even worse.

"A lot of people just want to let it go up there [in Appalachia] and let it rip," he says, his voice rising. "They say it'll act a lot like the northern hemlock, this, that, and the other thing. Yeah, maybe. When it comes to introducing nonnative species, we have such overwhelming evidence of good ideas gone bad . . . and this isn't just the Conservancy's policy, it's my personal policy . . . it's very dangerous tinkering."

This is the longstanding conservation credo: with enough space, money, and knowledge, we can protect natural places and, in many cases, restore them by stitching them back together. But while we're welcome to restore, redesign is frowned upon; *that* sort of tinkering crosses an invisible line between humans and capital-N Nature and risks making things much worse. We've good reason to distrust ourselves, after all. Until the 1950s, we thought planting kudzu was a good idea.

But climate change calls all this into question. If rising temperatures and changing weather patterns make restoration difficult or impossible, new brands of meddling may sometimes be the only alternative to extinction. Connie Barlow believes *Torreya taxifolia*, with its almost absurdly gloomy prospects in its current range, already requires a new strategy — and she welcomes the chance to provide it.

Barlow describes herself as "more interventionist" than many of the scientists and conservationists she encounters, explaining

that her background in ecology and evolutionary biology have im-
mersed her in the long time scales of evolution. "I don't have a
sense of what's normal," she says. "I do have a sense of species moving a lot through time."

Following her first visits to Torreya State Park, Barlow started an
e-mail correspondence with botanists, conservationists, and others
about the future of the tree. Some, such as paleoecologist Paul
Martin, loved the idea of moving *T. taxifolia* north. The Florida
torreya is widely believed to be an ice-age relict, "left behind" after
the last glacial retreat and very possibly better suited for cooler climates, with or without global warming. So why not return it to
the southern Appalachians, where it grew during the Pleistocene?
These arguments were countered by an ecologist named Mark
Schwartz, who has studied the Florida torreya at the Apalachicola Bluffs preserve since the late 1980s, and remains one of the
scant handful of scientists with in-depth knowledge of the species.
Schwartz defended the chances for restoration in the species' pres-
ent-day range. Before long the discussion reached an impasse, and
the disagreement found an audience.

In a 2004 forum in the now-defunct journal *Wild Earth,* Barlow
and Martin made what might be the first public case for assisted migration. Moving even federally endangered plants like the Florida
torreya to more favorable climates, they wrote, was "easy, legal, and
cheap," and *Torreya taxifolia,* prevented by highways, topography,
and its own biology from moving quickly on its own, needed immediate help.

While horticulturists at the Atlanta Botanical Garden have spent
years raising *Torreya taxifolia* in greenhouses and seminatural "potted orchards" in northern Georgia, Barlow and Martin dismissed
these efforts, saying that "potted is the botanical equivalent of
caged." They proposed that *T. taxifolia* be planted on privately
owned forest lands in southern Appalachia, easily four hundred
miles from the Florida Panhandle. The risk of the slow-growing,
problem-prone Florida torreya becoming an invasive weed is vanishingly small, they argued, and in the Appalachian forests, the
tree might even take the place of the eastern hemlock, another
subcanopy conifer in precipitous decline.

Schwartz, now a professor at the University of California, Davis, responded by acknowledging both the critical situation of the

Florida torreya and the possibility of healthier habitat in Appalachia. But he balked at assisted migration for much the same reasons that David Printiss — and many conservationists of all stripes — meet the idea with almost visceral hostility. The Florida torreya is unlikely to become the next kudzu, but the next species on the poleward wagon might very well prove a nasty invasive. And since scientists don't know precisely what climate change will mean for *Torreya taxifolia* and other species, conservationists can only make rough predictions about future habitats and future relationships among species. The unknowns are staggering.

If the theory of assisted migration isn't controversial enough, Schwartz points out, the reality is sure to be even more contentious: while people may be willing to export familiar species to safer habitats, they're less likely to open their home ecosystems to exotic refugees. "Here in northern California, if we were to ask people whether we can move a salamander that's going extinct because of climate change into Oregon, people would probably say yes," Schwartz says. "But if we ask people whether we can introduce a southern California species into a redwood grove for the same reason, they would uniformly say no way!"

Perhaps the most disturbing implication of assisted migration is that the traditional conservation notion — call it an illusion, if you like — of a place to get back to will disappear for good. Yet with or without assisted migration, that pristine place is already slipping out of reach. The demarcation between managed and wild has always been tenuous, defined more by emotion than data, and weakened over decades by the global reach of humankind: acid rain, DDT, PCBs, the traces of Prozac in rivers and streams. Climate change is the most dramatic transgression yet, for its effects range from pole to pole and can't be fenced in, mopped up, or halted by a National Park Service boundary.

Climate change is altering the wilderness peak, the back-yard nature preserve, the wild and scenic desert river — all the longstanding conservation victories, the places that not only lend inspiration and solace to the conservation movement but also prove the wisdom of its tactics. In transforming places once thought protected, in violating hard-fought boundaries, climate change is busting the limits of conservation itself.

*

The passionate critics of assisted migration didn't stop Connie Barlow, who moved briskly ahead with her plans for the Florida torreya. She created a Web site called the Torreya Guardians, where she and a handful of amateur horticulturists began to trade information about *Torreya taxifolia* cultivation in other habitats.

Their vision of the Florida torreya's future begins in the mountains of north Georgia, where the roads narrow and twist, and travel is measured in time instead of distance. Here Jack Johnston, a sleepy-eyed emergency-room nurse and amateur horticulturist, started growing Florida torreya after meeting Connie Barlow at a dinner in North Carolina. On the steep ground behind his house, on terraces that legend has it were used for growing corn to make white lightning in the 1930s, Johnston is cultivating a half-dozen *Torreya taxifolia* seedlings he bought, legally, from a nursery in South Carolina. Each is about two feet high, five years old, and healthy.

Johnston, whose isolated property is full of other rare plants ("I'm moving all sorts of things north," he jokes) is pleased by the apparent flexibility of his charges and nonchalant about the implications of assisted migration. "People have been moving plants around for a long time," he says. "This idea that we should be territorial about our plants, well, that's just kind of a provincial attitude."

The next day, during a long-awaited rainstorm in western North Carolina, Lee Barnes, the de facto lieutenant of the Torreya Guardians, is eager to talk *Torreya*. "I'm a horticulturist," he says. "I'm a professional tinkerer." Barnes, who is no stranger to *T. taxifolia* — he wrote his doctoral dissertation on the cultivation of the Florida torreya and two other endangered Florida species in the 1980s — has so far collected and distributed about 120 seeds to about a dozen people and gardens north of Georgia, including amateur gardeners in Ohio, New York, England, Switzerland, and elsewhere. Some recipients have reported their successes and failures; some have not.

Barnes's seed supply comes from a single grove of *Torreya taxifolia,* which grows not in Florida but about thirty miles from his home in North Carolina. In the 1930s and 1940s, on the grounds of George Vanderbilt's grand Biltmore estate, an enterprising head gardener planted seeds he and his botanical accomplices (known as the Azalea Hunters) collected throughout the Southeast. Today

lines of tourists snake through the vast gardens, but few notice the
unassuming, thin-limbed conifers that stand, unmarked, among
magnolias, pines, oaks, and redwoods.

Bill Alexander, forest historian for the estate, has lived on these
grounds for twenty years, and he walks along the curving path
through this cultivated forest, pointing out each Florida torreya in
turn. These trees, all apparently free of the disease that scourges
the Panhandle populations, were likely planted in the 1930s or
1940s — though perhaps as early as the 1890s — and some reach
fifty feet, a height now unimaginable in Florida. Despite freezes
and hurricanes, the Florida torreya has done itself proud in North
Carolina: one of the trees at Biltmore, Alexander believes, is the
second-largest of the species. The largest stands on a farm in northeastern North Carolina, surrounded by rusting farm equipment.

Alexander, who traces his family back to some of the first European settlers in the Biltmore area, is no ecosaboteur, but he likes
the democratic, do-it-yourself approach of the Torreya Guardians,
and he wants to see the species survive, no matter its longitude and
latitude. He says he'll happily supply seeds to the group as long
as the Biltmore trees continue to produce. And if the resulting
seedlings establish themselves outside gardens and the manicured
grounds of the Biltmore estate? Alexander looks pleased. "Well,"
he says, "then I'll think, 'By God, we've been successful.'"

In 2007, ecologist Mark Schwartz and two colleagues, Jessica Hellmann and Jason McLachlan, published a paper that modestly proposes a "framework for debate" on assisted migration. While they
criticized "maverick, unsupervised translocation efforts," such as
the Torreya Guardians', for their potential to undermine conservation work and create conflict, they directed their harshest criticism
at "the far more ubiquitous 'business as usual' scenario that is
the current de facto policy." The three scientists take different
stands on the notion of assisted migration. All are cautious, but
McLachlan is usually the most skeptical, and Hellmann, a University of Notre Dame ecologist who studies butterflies on the northern end of their range in British Columbia, is the most open to the
concept. "It's incredibly exciting to think that we could come up
with a strategy that might help mitigate the impacts of climate
change," she says.

Last fall, to initiate a broader discussion, the three scientists organized a meeting in Davis, California, with other researchers, land managers, environmental groups, and even an environmental ethicist. The Florida torreya isn't the only species that might benefit from immediate assisted migration. The Quino checkerspot butterfly has blinked out on the southern end of its range, in the Mexican state of Baja California, while the northern end of its range, in southern California, has been transformed by development. In South Africa and Namibia, rising temperatures on the northern edge of the range of the quiver tree are killing the succulent plants before the species has a chance to shift south.

But assisted migration is in no case a clear solution. Beyond initial concerns about new invasive species and territorial conflicts among conservationists, the meeting in California raised new questions. What if assisted migration is used to justify new habitat destruction? Who decides which species are moved, and who moves them? Isn't "assisted colonization" a more appropriate name than "assisted migration," which reminds people of birds on the wing?

Some researchers also worry that continued discussion about the strategy — which most agree is a last resort, likely too expensive and complicated for widespread use — distracts from the more prosaic, immediate duties of conservation and restoration. The Brown University ecologist Dov Sax, an invasive-species researcher working on assisted migration, has grander hopes for the conversation. "Conservation has really been built around a static view of the world," he says. "Given that climate change is going to happen, we need a whole new suite of strategies that could complement the old ones. This could get more people thinking about the other strategies we need."

Discussions of climate change always seem to end with a dreary litany of required sacrifices, uncomfortable changes that will be demanded of the penitent. There is no doubt that stabilizing the climate will require deep, society-wide reforms, some of them costly. But as climate change delivers its inconvenient truths, it also asks us to chuck a persistent and not-very-useful notion: the idea that conservation, and by extension restoration, is about gilt-framed landscapes.

Commitment to particular places and their histories has taken conservation a long way. It gives conservationists ground to stand on, in ways that range from the literal to the spiritual to the political. And restoring these beloved places to past states can restart ecological processes still relevant to the present day. But this sort of restoration works only when the climate is more or less stable — when the past supplies a reasonable facsimile of the future. Restoration ecologists remind us that the most effective restoration focuses not on a given point in the past but on the revival of clogged or absent natural processes. When climate change makes historical analogues irrelevant, it's these processes that will help species and systems survive in a new world.

Don Falk, an ecologist at the University of Arizona and the first executive director of the Society for Ecological Restoration, argues that assisted migration is simply another way to impersonate the process of dispersal: its adherents intend to transport species from places humans have made uninhabitable, through places humans have made impassable. Despite its undeniable risks, it may not be as radical as it first seems. It may be just another step in the evolution of conservation.

The job is no longer — if it ever was — to fence off surviving shards of landscape or to try to put everything back the way it used to be. Climate change requires conservationists to husband not a fixed image of a place but instead the fires, floods, and behaviors that create it, in order to help species and natural systems respond to a host of changes we're only beginning to understand. Assisted migration is certainly not the right strategy for all species — and given its myriad possible pitfalls, it may not be the right choice for any species. Yet the idea of it, and the discussion it provokes, point toward the future.

Mark Schwartz, for his part, still holds out hope for the recovery of the Florida torreya in Florida, for a small but healthy population of trees in the shady steephead ravines. But each time he visits the Panhandle, he says, he sees fewer and fewer *Torreya taxifolia*.

BENJAMIN PHELAN

How We Evolve

FROM *Seed*

WHEN THE PREVIOUS GENERATION of life scientists was coming up through the academy, there was a widespread assumption, not always articulated by professors, that human evolution had all but stopped. It had certainly shaped our prehuman ancestors — *Australopithecus, Paranthropus,* and the rest of the ape-men and man-apes in our bushy lineage — but once *Homo sapiens* developed agriculture and language, it was thought, we stopped changing. It was as though, having achieved its aim by the seventh day, evolution rested. "That was the stereotype that I learned," says the population geneticist and anthropologist Henry Harpending. "We showed up 45,000 years ago and haven't changed since then."

The idea makes a rough-and-ready kind of sense. Natural selection derives its power to transform from the survival of some and the demise of others and from differential reproductive success. But we nurse our sick back to health, and mating is no longer a privilege that males beat each other senseless to secure. As a result, even the less fit get to pass on their genes. Promiscuity and sperm competition have given way to spiritual love; the fittest and the unfit are treated as equals and equally flourish. With the advent of culture and our fine sensibilities, the assumption was, natural selection went by the board.

Moreover, evolution had never been observed in humans, except in a few odd cases, so the conclusion was drawn that it wasn't happening. One can't fault the logic. The most famous case of adaptive change in humans, that of sickle cell trait as an evolutionary response to malaria, seemed to prove the point that human evolution must be rare: even in as dire and malaria-stricken an environment as West Africa, the only response evolution has been able to come up with is an imperfect defense that can cause serious health problems along with its solitary benefit. Selection pressures as strong as those brought about by endemic malaria are uncommon, and civilization was thought to wash out those less powerful.

But since the turn of the millennium, genomics has undergone a revolution. With the completion of such landmark studies as the Human Genome Project and the publication of HapMap, scientists finally have access to the particles of evolution. They can inspect vast stretches of DNA from people of all ethnicities, and the colossal amount of information suddenly available has spurred a revision of the old static picture that will render it unrecognizable. Harpending and a host of researchers have discovered in our DNA evidence that culture, far from halting evolution, appears to accelerate it.

John Hawks started out as a "fossil guy" studying under Milford Wolpoff, a paleoanthropologist who is the leading proponent of the faintly heretical multiregional theory of human evolution. Coming to genetics from such a background has perhaps given Hawks the stomach to wield unfashionable hypotheses. In December 2007 he, Harpending, and others published a paper whose central finding, that evolution in humans is observable and accelerating, would have been nonsensical to many geneticists twenty years ago. Up to 10 percent of the human genome appears to be evolving at the maximum rate, more quickly than ever before in human history.

"Seven percent is a minimum," Hawks says. "It's an amazing number," and one that is difficult to square with the prevailing view of natural selection's power. Because most mutations have a neutral effect on their carriers, making them neither fitter nor less fit, neither more fertile nor sterile, only slightly different, those changes are invisible to natural selection. They spread or don't spread through a population by chance, in a process called genetic drift, which is often thought of as the agent of more change than natural selection. But the changes that Hawks detected, if he is correct, are too consistent from person to person, from nationality to nationality, to have been caused by genetic drift alone.

By looking at the data from HapMap, a massive survey of the genetic differences between selected populations from around the world, Hawks identified gene variants, or alleles, that were present in many people's DNA but not in everyone's. These alleles seemed to be moving, over time, through populations in a way that matched mathematical predictions of what natural selection should look like on the genomic level. And though Hawks doesn't know why possession of the new alleles should be advantageous, he doesn't need to know. The signature that natural selection inscribes on the genome is legible even when the import of the message is unclear.

HapMap can reveal where natural selection has occurred thanks to the tendency of DNA "neighborhoods" to be inherited in blocks that do not change much, if at all, from parent to offspring. When an organism reproduces, adjacent DNA sticks together and is passed on as a unit to the offspring. Such sections of linked DNA are called haplotypes; HapMap is a directory of them. A given haplotype is nearly identical among family members, but populations that have had recent contact with each other, such as the French and the Spanish, or the Cherokee and the Inuit, also tend to share it.

One of the characteristics of this linkage is that it is strong over short distances on a chromosome and weak over long distances. This is because mutations are rare but equally likely at every location, so they happen less often in a small region of DNA than in a large region. Over many generations, mutations nibble away at the edges of haplotypes and poke holes in their interiors, and the routine reshuffling of nucleotides, called recombination, can move linked sections of DNA far from one another, thereby breaking the linkage. Thus the length of a haplotype roughly indicates its age, as does the amount of variation within it. Since mutations and recombination occur at a predictable rate, by comparing haplotypes from two populations, one can determine their degree of relatedness and thus estimate how long ago they diverged. So, for example, the San of southern Africa and the Han Chinese would tend not to share haplotypes because their populations diverged long ago, and those they did share would be short or contain a great deal of variation.

Because haplotypes are similar from population to population, differences are easy to spot; a variation on a familiar haplotype background is like a smear of red paint on a white wall. If one looks at a haplotype in one hundred individuals, and ninety of them are identical but ten show the exact same variation, the odds are vanishingly small that random processes generated the same mutation ten different times. Such a site is a candidate for one undergoing natural selection, because only a mutation that confers some kind of advantage will be propagated reliably through the population.

If the trait under selection produces a significant enough adaptive advantage, the allele responsible for the trait will rise in frequency so quickly that it will drag a long haplotype along with it before recombination and mutation can break it down to a short haplotype. So a rare allele on a long haplotype is an indication of strong and recent selection.

Hawks's analysis of the HapMap data yielded many such candidate sites, but some of his colleagues were unimpressed. "They didn't like the idea," he says. An anonymous reviewer of his paper claimed not to think that natural selection could possibly be important in recent evolution, "so much so that they said positive selection happens rarely, if ever."

An oft-cited example of evolution in historic times is the spread of the mutation that allows humans to digest milk in adulthood. It seems to have arisen around eight thousand years ago and has since spread to all parts of the world, though there are still plenty of us without it: one in fifty Swedes and nine out of ten Asian Americans lack the mutation. The lactose intolerant are, at least in this respect, like the first *Homo sapiens*.

"There are five versions of the lactase drinking gene, so five different populations have mutations that let them drink milk," says Hawks. Because of this, many mutations conferring the same benefit are unlikely to have become common by genetic drift, but Hawks knows of practicing geneticists who find the idea that natural selection was the agent of their propagation to be preposterous. He's incredulous at what he sees as such scientists' fundamental misprision of the field's core principle. "This is Darwin's field," he says. "Darwin talks about evolution, and Darwinism is about natural selection. But these people don't believe in natural selection — except way back when, when chimps and humans were the same."

By Hawks's own description, his research "depends on a view of evolution that's dominated by natural selection. When I look at the evolutionary processes leading to humans I'm thinking, what's the adaptive change that's happening? What are the constraints on our adaptation?" One of those, he says, "is demography."

By invoking ancient demography via the anthropological record, Hawks believes he has identified what has been driving all the adaptive evolution he detected: an explosion in the global human population roughly coincident with the agricultural revolution of some ten thousand years ago. We invented agriculture, started eating different food, and began dwelling in cities. Our numbers swelled, our world changed, and our DNA is still catching up.

Spencer Wells, director of the Genographic Project, an attempt to reconstruct human migration patterns by sampling DNA from the world's populations, has studied humanity's transition to agriculture extensively. Hawks's result was no surprise to him.

"The biggest change in our lifestyle as a species has happened in the past ten thousand years," Wells says. "We spent most of the past million or so years of evolution living as hunter-gatherers, hunting game on the African savannas, or gathering shellfish on the coast, gradually moving out to Eurasia. Then suddenly, in the past ten thousand years, we become a species that settles down. The diversity of food sources drops precipitously from over one hundred in the hunter-gatherer diet to fewer than ten in the average agricultural diet. And then, of course, you build up the population densities, and disease takes off."

Such changes to environment, diet, and disease load are classic agents of natural selection. The three acting in concert could certainly accelerate evolution. But it might seem odd that a larger population is required to produce a faster rate of evolution, especially if you happen to be American.

The early- and mid-twentieth century witnessed a tension between two interpretations of evolutionary theory. Sewall Wright, an American, argued that for rapid evolution to occur, what was required was a small, semi-isolated population through which a mutation could spread quickly, even by genetic drift. Thereafter, that population could migrate and spread the allele in other populations. But R. A. Fisher, a Brit, argued that in fact a large population was required, because only a large population can produce large numbers of mutations. Because most mutations are neutral, he reasoned, it takes a large number of mutations to produce one beneficial allele. American biologists were most influenced by Wright, but Fisher's work is where Hawks and Harpending find their support.

Fisher developed a mathematical model of how beneficial mutations should move through a population toward fixation, the point at which all members of a species have the allele. The shape of the curve is characterized by slow dispersion at first, because the mutation initially exists in only one member of the species. It takes a long time for a new allele to reach an appreciable frequency in a population, but at a certain point the growth rate becomes much steeper; many carriers bear many offspring, and the gene becomes widespread. But during the last leg of the push toward fixation, the rate decreases and begins to resemble a curve approaching an asymptote.

When anthropologists analyzed caches of ancient Eurasian skeletons, they found evidence that Fisher's model was correct. In the DNA of a group of five-thousand-year-old skeletons from Germany, they discovered no trace of the lactase allele, even though it had originated a good three thousand years beforehand. Similar tests done on three-thousand-year-old skeletons from Ukraine showed a 30 percent frequency of the allele. In the modern populations of both locales, the frequency is around 90 percent.

"This is the curve that Fisher predicted," says Hawks. "The frequency [of the lactase allele] that we have at different times hits this curve. This means that the maximum rate of change in frequency of this gene was within the past three thousand years, even though the gene originated eight thousand years ago."

Seeing the mathematical model he was using borne out in data other than his own was encouraging to Hawks: many of the alleles he'd identified as being under selection seem to show a similar trajectory toward fixation.

"My attitude about recent human evolution comes straight out of mathematics," he says. "I can say, this is population growth, and these are the effects it should have. And as long as I keep observing data that are consistent with that idea, I think it's a strong model . . . Once you can connect history with genes, you can build up knowledge from the standpoint of anthropology, then let the biochemists work out what each gene does."

Being able to understand the purpose of a given gene, however, is perhaps the main challenge facing the current generation.

Hawks doesn't know what function the genes he identified as evolving perform, but such information isn't important for his purposes. He is content with linking demographic history with mathematics and gene surveys and hypothesizing natural selection based on the confluence of those streams of evidence. A biochemist, though, might balk at saying that a gene is under selection without knowing what the gene actually does.

"Human genetics made a major leap forward at the turn of the millennium," says Pardis Sabeti, an evolutionary geneticist at MIT's Broad Institute who has done a great deal of work on methods for assessing genomic surveys like HapMap, the first draft of which was published in 2005. HapMap is a leaner and in some ways more powerful version of the Human Genome Project, as it compiles only those regions of the human genome — less than 1 percent — that have the potential to differ from person to person. In comparing different populations' genetic information, it's possible to tease out patterns of gene inheritance, how certain genes correlate with certain diseases, and even the likely geographic origin of some mutations.

One of the methods that Sabeti has developed to identify selection is to search for rare alleles on long haplotypes, which is useful for identifying selection in the past thirty thousand years or so. Using the long-haplotype test on HapMap data, Sabeti was able to find what appears to be a signature for recent natural selection on genes that are associated with resistance to Lassa, a hemorrhagic fever that's endemic to parts of central and western Africa. She is perhaps more cautious than Hawks in her conclusions, though; they are in different fields and have different standards of proof.

"I'm a little guarded on the findings for Lassa, because the question is, is the finding real?" she says. "The strongest signal of selection we've detected in a West African population is on a gene called *Large,* which has been biologically linked to Lassa." Lassa is a poorly understood and infrequently studied pathogen, she says, so there was not much literature to consult about genes possibly associated with it. However, a microbiologist named Stefan Kunz had demonstrated that if *Large* is deleted from a mouse's DNA, Lassa is unable to infect it.

"That was exciting, because otherwise we'd look at the gene and say 'I don't know what it does,' and that would have been the end of it. But now we could see a link," she says. "But when you look at selection you never believe your results completely because it's circumstantial. We have basic evidence that it seems to be evolving and we can link it to this disease, but we don't have a real biological link."

The molecular record, for all its overwhelming garrulousness, its babel of A's, C's, G's, and T's, is ambiguous. But the fossilized skulls of our ape lineage seem to tell a clear story with respect to one trait, anyway. The past few million years have witnessed a steady, plodding increase in the volume of the human lineage's brains and, presumably, the sophistication of their contents. High intelligence is to great apes as the wing is to birds.

But where are we in that process? Is intelligence still being selected for? Parsimony and uniformitarianism would compel one to answer yes; things in the present are, by and large, as they were in the past. But the way evolution works, whereby mutations arise in one person and slowly spread throughout a population, makes such a question difficult to frame, for if intelligence is still under selection, that could mean that some populations at this very moment are slightly smarter than others — that, perhaps, even certain ethnicities are slightly smarter than others. In the West, speculation on the subject almost automatically tars the speculator as a eugenicist or a racialist.

Bruce Lahn is an evolutionary geneticist and a lab director at the University of Chicago, but he was born and completed the early part of his education in China. A heightened sensitivity to imputations of racialism doesn't afflict most Chinese, according to Lahn, who, by his own admission, has yet to fully internalize the finer points of Western political correctness. In a pair of 2005 papers, he presented evidence that two genes known to play a role in brain development, *microcephalin* and *ASPM,* appear to be undergoing continuing natural selection in historical times. In the penultimate paragraph on *microcephalin,* he observed that "Sub-Saharan populations generally [have] lower frequencies than others." And after noting that the *ASPM* mutation, which he refers to as haplogroup D, is most common in Europeans and Middle Easterners and least common in sub-Saharan Africans, he speculated, "Although the age of haplogroup D and its geographic distribution across Eurasia

roughly coincide with . . . the development of cities and written language 5,000 to 6,000 years ago around the Middle East, the significance of this correlation is not yet clear."

It makes sense that some alleles present in Europe, Asia, and the
rest of the world wouldn't appear in sub-Saharan Africa, and vice
versa; population flow has not yet had time to spread all alleles to
all parts of the world. However, it's hard for many of us not to hear
in Lahn's musings on brain genes the ugly implication that Africans are inferior. But such was not Lahn's intention, nor was that
his finding. It was not even what he was investigating.

"Some interpret it as meaning this is the civilization gene, which
is clearly not what we're trying to say. Maybe we should have said
it with more qualifications, to avoid the misconception," he says.
The belief that minor mutations to two genes could bring about a
profound and essential difference to an abstract quality as poly-
morphous as intelligence Lahn sees as springing from America's
confusion about race, its desire to overcome a shameful past, and
a fear that old racist beliefs might be given empirical support. Nevertheless, Lahn and his group did ultimately investigate whether
possession of the new alleles correlated with intelligence. It did
not.

Indeed, possession of Lahn's variants might have nothing to do
with intelligence. "It could impact emotionality, the ability to be patient, for example," he says. "Our understanding of brain evolution
at the phenotype level is so rudimentary right now. We're very far
from actually breaking down the difference between human and
other species, let alone among humans."

Lahn's result was criticized in subsequent papers, not on ideological grounds but on technical ones. It was claimed that the signal for selection he thought he'd found was not there. He took the
criticism in stride and reanalyzed his data. "We stand by our conclusions," he says. "We have more unpublished data to support them.
We're convinced that what we published is real."

Even if Lahn could prove to everyone's satisfaction that *ASPM*
and *microcephalin* are under selection, whether intelligence is the
trait being selected for would be far from a settled question. It
could be, as Lahn suggested, that some other mental trait is being
selected or that the activity of *ASPM* and *microcephalin* in other parts
of the body is what is under selection. More work will certainly be
done. But one can speculate with far more confidence about what

drove the dramatic increase in intelligence attested by the fossil record: the advent of human culture.

"Intelligence builds on top of intelligence," says Lahn. "[Culture] creates a stringent selection regime for enhanced intelligence. This is a positive feedback loop, I would think." Increasing intelligence increases the complexity of culture, which pressures intelligence levels to rise, which creates a more complex culture, and so on. Culture is not an escape from conditioning environments. It is an environment of a different kind.

Lahn says there could be "some deep-down information theory perspective" that underlies both the rapid increase in human intelligence and an event like the Cambrian explosion, the unequaled diversification of life forms that occurred about 500 million years ago. In an eye blink, almost every modern body plan came into existence. "It may take a long time to evolve certain components of the body plan, but once you have them, minor tinkering that requires not many changes and very little evolutionary time could give you great diversity in body plans and species," Lahn says. "The brain may be similar, because it takes a long time to get to a certain level of intelligence, but once you get there, it makes possible a cultural explosion."

Both events inched toward a threshold that, once crossed, was soon left far behind. The twentieth century, in which it took us a mere sixty years to elaborate the horse-drawn carriage into a vehicle that carried us to the moon, and the howitzer into a fifty-megaton nuclear weapon, was another threshold. The forces that we created are on a different scale from those of nature, which works slowly. It seems possible that as our technology grows more subtle, genetic manipulation, gene manufacturing, and even cloning could finally carry us clear of natural selection, but such a commanding position can be maintained only with the survival of a technological society, and that is hardly a foregone conclusion.

The bleakness of that vision exerts a strong hold on Paul Ehrlich, a professor of population studies at Stanford, who finds in the twentieth century a minefield of near misses with extinction. We were saved as often by cunning as by dumb luck: intended to save sleeping families from exploding refrigerators cooled by ammonia, chlorofluorocarbons nearly fried the entire planet. As often as not, some solution creates a new problem.

"The fate of our civilization, and maybe our species," says Ehrlich, "may be determined by the next five generations. So I don't really give a shit what's happening to our genetic evolution." The global climate is changing too violently for DNA to respond by fiddling around with heat regulation and hair thickness; forests everywhere are being clear-cut too quickly for their inhabitants to adjust, so food chains are coming undone; the collapse of global fisheries has been identified as an imminent calamity; and a nuclear disaster would constitute a catastrophe many orders of magnitude larger than what nature could readily absorb. If any of these nightmare scenarios comes to pass, Ehrlich fears, evolution will be unable to help us. It may be operating faster than we thought, but it's not that fast. Problems like smog and acid rain seem almost quaint, and even to be longed for.

Species are transient. There is no question that the day will come when humans are no longer on Earth. But the transience to which we are subject has two faces. The first is extinction. Unlike our forebears, we are aware of how tenuous our perch atop the food chain is. It remains to be seen whether that knowledge has been acquired too late to be of use.

The second face of *Homo sapiens'* eventual exit from history is the more hopeful possibility that we may yet evolve into our own successors. Unlike our forebears, we are aware of evolution, which changes our relationship to it, if only by a little, for we are still natural creatures. We continue to evolve in the face of hunger, disease, and a changing ecosystem; but our virtual habitat of culture could enable us to become both subjects of evolution and conscious codirectors of it. "It's occurring," says Ehrlich. "There's no question about it. What's frightening is the questions we'll have to ask."

Science must evolve new tools to raise us to such a commanding vantage, as well as to avert a self-inflicted extinction. Technology might someday enable us to control aspects of evolution, or it may prove to be the ultimate selection regime, culling all of us. Perhaps we already find ourselves wishing we'd lacked the intelligence to monkey with howitzers. Either way, the culture that we've created is, strangely, evolution's most powerful tool and its potential nemesis, the womb of human nature and perhaps its grave. By our own hand: this is how we evolve.

VIRGINIA POSTREL

Pop Psychology

FROM *The Atlantic Monthly*

IN THESE UNCERTAIN ECONOMIC TIMES, we'd all like a guar
anteed investment. Here's one: it pays a 24-cent dividend every
four weeks for sixty weeks, fifteen dividends in all. Then it disappears. Unlike a bond, this security has no redemption value. It simply provides guaranteed dividends. It involves no tricky derivatives
or unknown risks. And it carries absolutely no danger of default.
What would you pay for it?

Before financially sophisticated readers drag out their calcula-
tors, look up interest rates, and compute the present value of those
future payments, I have a confession to make. You can't buy this security, and it doesn't really pay dividends every four weeks. It pays
every four *minutes,* in a computer lab, to volunteers in economic
experiments.

For more than two decades, economists have been running versions of the same experiment. They take a bunch of volunteers,
usually undergraduates but sometimes businesspeople or graduate students; divide them into experimental groups of roughly a
dozen; give each person money and shares to trade with; and pay
dividends of 24 cents at the end of each of fifteen rounds, each lasting a few minutes. (Sometimes the 24 cents is a flat amount; more
often there's an equal chance of getting 0, 8, 28, or 60 cents, which
averages out to 24 cents.) All participants are given the same information, but they can't talk to one another, and they interact
only through their trading screens. Then the researchers watch
what happens, repeating the same experiment with different small
groups to get a larger picture.

The great thing about a laboratory experiment is that you can control the environment. Wall Street securities carry uncertainties — more, lately, than many people expected — but this experimental security is a sure thing. "The fundamental value is unambiguously defined," says the economist Charles Noussair, a professor at Tilburg University in the Netherlands, who has run many of these experiments. "It's the expected value of the future dividend stream at any given time": 15 times 24 cents, or $3.60 at the end of the first round; 14 times 24 cents, or $3.36 at the end of the second; $3.12 at the end of the third; and so on down to zero. Participants don't even have to do the math. They can see the total expected dividends on their computer screens.

Here, finally, is a security with security — no doubt about its true value, no hidden risks, no crazy ups and downs, no bubbles and panics. The trading price should stick close to the expected value.

At least that's what economists would have thought before Vernon Smith, who won a 2002 Nobel Prize for developing experimental economics, first ran the test in the mid-1980s. But that's not what happens. Again and again, in experiment after experiment, the trading price runs up way above fundamental value. Then, as the fifteenth round nears, it crashes. The problem doesn't seem to be that participants are bored and fooling around. The difference between a good trading performance and a bad one is about eighty dollars for a three-hour session, enough to motivate cash-strapped students to do their best. Besides, Noussair emphasizes, "You don't just get random noise. You get bubbles and crashes" — 90 percent of the time.

So much for security.

These lab results should give pause not only to people who believe in efficient markets but also to those who think we can banish bubbles simply by curbing corruption and imposing more regulation. Asset markets, it seems, suffer from irrepressible effervescence. Bubbles happen, even in the most controlled conditions.

Experimental bubbles are particularly surprising because in laboratory markets that mimic the production of goods and services, prices rise and fall as economic theory predicts, reaching a neat equilibrium where supply meets demand. But, like real-world purchasers of haircuts or refrigerators, buyers in those markets need to know only how much they themselves value the good. If the price is less than the value to you, you buy. If not, you don't, and vice versa for sellers.

Financial assets, whether in the lab or the real world, are trickier to judge: can I flip this security to a buyer who will pay more than I think it's worth? In an experimental market, where the value of the security is clearly specified, "worth" shouldn't vary with taste, cash needs, or risk calculations. Based on future dividends, you know for sure that the security's current value is, say, $3.12. But — here's the wrinkle — you don't know that I'm as savvy as you are. Maybe I'm confused. Even if I'm not, you don't know whether I know that you know it's worth $3.12. Besides, as long as a clueless greater fool who might pay $3.50 is out there, we smart people may decide to pay $3.25 in the hope of making a profit. It doesn't matter that we know the security is worth $3.12. For the price to track the fundamental value, says Noussair, "everybody has to know that everybody knows that everybody is rational." That's rarely the case. Rather, "if you put people in asset markets, the first thing they do is not try to figure out the fundamental value. They try to buy low and sell high." That speculation creates a bubble.

In fact, the people who make the most money in these experiments aren't the ones who stick to fundamentals. They're the speculators who buy a lot at the beginning and sell midway through, taking advantage of "momentum traders," who jump in when the market is going up, don't sell until it's going down, and wind up with the least money at the end. ("I have a lot of relatives and friends who are momentum traders," comments Noussair.) Bubbles start to pop when the momentum traders run out of money and can no longer push prices up.

But people do learn. By the third time the same group goes through a fifteen-round market, the bubble usually disappears. Everybody knows what the security is worth and realizes that everybody else knows the same thing. Or at least that's what economists assumed was happening. But work that Noussair and his coauthors published in the December 2007 *American Economic Review* suggests that traders don't reason that way.

In this version of the experiment, participants took part in the fifteen-round market four times in a row. Before each session, the researchers asked the traders what they thought would happen to prices. The first time, participants didn't expect a bubble, but in later markets they did. With each successive session, however, they predicted that the bubble would peak later and reach a higher price than it actually did. Expecting the future to look like the past, they traded accordingly, selling earlier and at lower prices than in the previous session, hoping to realize a profit before the bubble burst. Those trades, of course, changed the market pattern. Prices were lower, and they peaked closer to the beginning of the session. By the fourth round, the price stuck close to the security's fundamental value — not because traders were going for the rational price but because they were trying to avoid getting caught in a bubble.

"Prices converge toward fundamentals ahead of beliefs," the economists conclude. Traders literally learn from experience, basing their expectations and behavior not on logical inference but on what has happened in the past. After enough rounds, markets work their way toward a stable price.

If experience eliminates bubbles in the lab, you might expect that more-experienced traders in the real world (or what experimental economists prefer to call "field markets") would produce fewer financial crises. When asset markets run into trouble, maybe it's because there are too many newbies: all those dot-com day traders, twenty-something house flippers, and newly minted MBAs. As Alan Greenspan told Congress in October 2008, "It was the failure to properly price such risky assets that precipitated the crisis." People didn't know what they were doing. What markets need are more old hands.

Alas, once again the situation is not so simple. Even experienced traders can make big mistakes when conditions change. In research published in the June 2008 *American Economic Review,* Vernon Smith and his collaborators first ran the standard experiment, putting groups through the fifteen-round market twice. Then the researchers changed three conditions: they mixed up the groups, so participants weren't trading with familiar faces; they increased the range of possible dividends, replacing four possible outcomes (0, 8, 28, or 60), averaging 24, with five (0, 1, 8, 28, or 98), averaging 27; finally, they doubled the amount of cash and halved the number of shares in the market. The participants then completed a third round. These changes were based on previous research showing that more cash and bigger dividend spreads exacerbate bubbles.

Sure enough, under the new conditions, the experienced traders generated a bubble just as big as if they'd never been in the lab. It didn't last quite as long, however, or involve as much volume. "Participants seem to be tacitly aware that there will be a crash," the economists write, "and consequently exit from the market (sell) earlier, causing the crash to start earlier." Even so, the price peaks far above the fundamental value. "Bubbles," the economists conclude, "are the funny and unpredictable phenomena that happen on the way to the 'rational' predicted equilibrium *if the environment is held constant long enough.*"

For those of us who invest our money outside the lab, this research carries two implications.

First, beware of markets with too much cash chasing too few good deals. When the Federal Reserve cuts interest rates, it effectively frees up more cash to buy financial instruments. When lenders lower down-payment requirements, they do the same for the housing market. All that cash encourages investment mistakes.

Second, big changes can turn even experienced traders into ignorant novices. Those changes could be the rise of new industries like the dot-coms of the 1990s or new derivative securities created by slicing up and repackaging mortgages. I asked the Caltech economist Charles Plott, one of the pioneers of experimental economics, whether the recent financial crisis might have come from this kind of inexperience. "I think that's a good thesis," he said. With so many new instruments, "it could be that the inexperienced heads are not people but the organizations themselves. The organizations haven't learned how to deal with the risk or identify the risk or understand the risk."

Here the bubble experiments meet up with another large body of experimental research, first developed by Plott and his collaborators. This work explores how speculative markets can pool information from lots of people ("the wisdom of crowds") and arrive at accurate predictions — for example, who's going to win the presidency or the World Series. These markets work, Plott explains, because people with good information rush in early, leading prices to reflect what they know and setting a trajectory that others follow. "It's a kind of cascade, a good cascade, just what should happen," he says. But sometimes the process "can go bananas" and create a bubble, usually when good information is scarce and people follow leaders who don't in fact know much.

That may be what happened on Wall Street, Plott suggests. "Now we have new instruments. We have 'leaders,' who one would ordinarily think know something, getting in there very aggressively and everybody cuing on them — as they have done in the past, and as markets should. But in this case, there might be a bubble." And when you have a bubble, you will get a crash.

DAVID QUAMMEN

Contagious Cancer

FROM *Harper's Magazine*

DURING THE EARLY MONTHS of 1996, not long before Easter, an amateur wildlife photographer named Christo Baars made his way to the Australian island-state of Tasmania, where he set up camp in an old airport shack within the boundaries of Mount William National Park. Baars's purpose, as on previous visits, was to photograph Tasmanian devils, piglet-size marsupials unique to the island's temperate forests and moors. Because devils are nocturnal, Baars equipped his blind with a cot, a couple of car batteries, and several strong spotlights. For bait he used road-kill kangaroos. Then he settled in to wait.

The devil, known to science as *Sarcophilus harrisii*, lives mostly by scavenging and sometimes by predation. It will eat, in addition to kangaroo meat, chickens, fish, frogs, kelp maggots, lambs, rats, snakes, wallabies, and the occasional rubber boot. It can consume nearly half its own body weight in under an hour, and yet — with its black fur and its trundling gait — it looks like an underfed bear cub. Fossil evidence shows that devils inhabited all of Australia until about five hundred years ago, when competition with dingoes and other factors caused them to die out everywhere but in Tasmania, which dingoes had yet to colonize. More recently, Tasmanian stockmen and farmers have persecuted devils with the same ferocity directed elsewhere at wolves and coyotes. The devils' reproductive rate, opportunistic habits, and tolerance for human proximity, however, have allowed localized populations to persist or recover, and at the time of Baars's 1996 visit, their total number was probably around 150,000.

On his earlier visits, Baars had seen at least ten devils every night, and they were quick to adjust to his presence. They would walk into his blind, into his tent, into his kitchen, and he could recognize returning individuals by the distinctively shaped white patches on their chests. This trip was different. On the first night, his bait failed to attract a single devil, and the second night was only a little better. He thought at first that maybe the stockmen and farmers had finally succeeded in wiping them out. Then he spotted a devil with a weird facial lump. It was an ugly mass, rounded and bulging, like a huge boil or a tumor. Baars took photographs. More devils wandered in, at least one of them with a similar growth, and Baars took more pictures. This was no longer wildlife photography of the picturesque sort; it was, or anyway soon would become, forensic documentation.

Back in Hobart, Tasmania's capital, Baars showed his pictures to Nick Mooney, a veteran officer of Tasmania's Parks and Wildlife Service, who has dealt with the devil and its enemies for decades. Mooney had never seen anything like this. The lumps looked tumorous, yes — but what sort of tumor? Mooney consulted a pathologist, who suggested that the devils might be afflicted with lymphosarcoma, a kind of lymphatic cancer, maybe caused by a virus passed to the devils from feral cats. Such a virus might also be passed from devil to devil, triggering cancer in each.

More evidence of contagion began to accumulate. Three years after Baars shot his photographs, a biologist named Menna Jones took note of a single tumor-bearing animal, something she had not seen before. Then, in 2001, at her study site along Tasmania's eastern coast, her traps yielded three more devils with ulcerated tumors. That really got her attention. She euthanized the animals and brought them to a lab, where they became the first victims to be autopsied by a veterinary pathologist. The "tumors" (until then the term had been only a guess or a metaphor) did seem to be cancerous malignancies, but not of the sort expected from a lymphosarcoma-triggering virus. This peculiarity raised more questions than it answered. Tasmanian devils in captivity were known to be quite susceptible to cancer, at least in some circumstances, possibly involving exposure to carcinogens. But the idea that the cancer itself was contagious seemed beyond the realm of possibility. And yet, during the following year, Menna Jones charted the spread of the problem across northern Tasmania. Nick Mooney, meanwhile, had done some further trapping himself. At a site in the northern midlands, he captured twenty-three devils, seven of which had horrible tumors. Shocked and puzzled, he remembered the Baars photos from years earlier.

Further trapping (more than a hundred animals, of which 15 percent were infected) showed Mooney what Jones had also seen: that the tumors were consistently localized on faces, filling eye sockets, distending cheeks, making it difficult for the animals to see or to eat. Why faces? Maybe because devils suffer many facial and mouth injuries — from chewing on brittle bones, from fighting with one another over food and breeding rights, from the rough interactions between male and female when they mate. The bigger tumors were crumbly, like feta cheese. Could it be that tumor cells, broken off one animal, fell into the wounds of another, took hold there, and grew? This prospect seemed outlandish, but the evidence was leading inexorably to a strange and frightening new hypothesis: the cancer itself had somehow become contagious.

Under ordinary circumstances, cancer is an individuated phenomenon. Its onset is determined partly by genetics, partly by environment, partly by entropy, partly by the remorseless tick-tock of time, and (almost) never by the transmission of some tumorous essence. It arises from within (usually) rather than being imposed from without. It pinpoints single victims (usually) rather than spreading through populations. Cancer might be triggered by a carcinogenic chemical, but it isn't itself poisoning. It might be triggered by a virus, but it isn't fundamentally viral. Cancer differs also from heart disease and cirrhosis and the other lethal forms of physiological breakdown; uncontrolled cell reproduction, not organ dilapidation, is the problem.

Such uncontrolled reproduction begins when a single cell accumulates enough mutations to activate certain growth-promoting genes (scientists call them oncogenes) and to inactivate certain protections (tumor suppressor genes) that are built into the genetic program of every animal and plant. The cell ignores instructions to limit its self-replication, and soon it becomes many cells, all of them similarly demented, all bent on self-replication, all heedless of duty and proportion and the larger weal of the organism.

That first cell is (almost always) a cell of the victim's own body. So cancer is reinvented from scratch on a case-by-case basis, and this individuation, this personalization, may be one of the reasons that it seems so frightening and solitary. But what makes it even more solitary for its victims is the idea, secretly comforting to others, that cancer is never contagious. That idea is axiomatic, at least in the popular consciousness. *Cancer is not an infectious disease.* And the axiom is (usually) correct. But there are exceptions. Those exceptions point toward a broader reality that scientists have begun to explore: cancers, like species, evolve. And one way they can evolve is toward the capacity to be transmitted between individuals.

Devil tumor isn't the only form of cancer ever to achieve such a feat. Other cases have occurred and are still occurring. The most notable is canine transmissible venereal tumor (CTVT), also called Sticker's sarcoma, a sexually transmitted malignancy in dogs. Again, this is not merely an infectious virus that tends to induce cancer. The tumor cells themselves are transmitted during sexual contact. CTVT is widespread (though not common) and has been claiming dogs around the world at least since a Russian veterinarian named M. A. Novinsky first noted it in 1876. The distinctively altered chromosome patterns shared by the cells of CTVT show the cancer's lineal continuity, its identity across space and through time. Tumor cells in Dog B, Dog C, Dog D, and Dog Z are more closely related to one another than those cells are to the dogs they respectively inhabit. In other words, CTVT can be conceptualized as a single creature, a parasite (and not a *species* of parasite, but an *individual*), which has managed to spread itself out among millions of different dogs. Research by molecular geneticists suggests that the tumor originated in a wolf, or maybe an East Asian dog, somewhere between 200 and 2,500 years ago, which means that CTVT is probably the oldest continuous lineage of mammal cells presently living on Earth. The dogs may be young, but the tumor is ancient.

Unlike devil tumor — now known as devil facial tumor disease, or DFTD — CTVT is generally not fatal. It can be cured with veterinary surgery or chemotherapy. In many cases, even without treatment, the dog's immune system eventually recognizes the CTVT as alien, attacks it, and clears it away, just as our own immune systems eventually rid us of warts.

The case of the Syrian hamster is more complicated. This tumor arose around 1960, when researchers at the National Cancer Institute, in Bethesda, Maryland, performed an experiment in which they harvested a naturally occurring sarcoma from one hamster and injected those cells (as cancer scientists often do) into healthy animals. When the injected hamsters developed malignancies, more cells were harvested. Each such inoculation-and-harvest cycle is called a passage. The experiment involved a dozen such passages, and over time the tumor began to change. It had evolved. The later generations, unlike the first, represented a sort of super tumor, capable of getting from hamster to hamster without benefit of a needle. The researchers caged ten healthy hamsters together with ten cancerous hamsters and found that nine of the healthy animals acquired tumors through social contact. The hamster tumor had leapt between animals — or anyway, it had been smeared, spat, bitten, and dribbled between them. (The tenth hamster was cannibalized before it could sicken.) In a related experiment, the tumor even passed between two hamsters separated by a wire screen. The scientists had in effect created a laboratory precursor of what would eventually afflict Tasmanian devils in the wild: a Frankenstein malignancy, a leaping tumor, which could conceivably kill off not just individuals but an entire species.

Early last summer I went to Tasmania, where I met Menna Jones for an excursion to the Forestier Peninsula, a long hook of land that juts southeastward into the Tasman Sea. Jones supervises an experimental trapping program aimed at ridding the peninsula of tumor disease or, at least, determining whether that goal is achievable. The Forestier is a good place for such trials because the peninsula (and its lower extension, a second lobe called the Tasman Peninsula) is connected to the rest of Tasmania by only a narrow neck — just a two-lane bridge across a canal. If the disease could be eradicated from the entire peninsula by removing all sick animals and leaving the healthy ones, Forestier and Tasman might be protected from reinfection by a devil-proof barrier across the bridge; and if that worked, the protected population could rebound quickly. The Forestier Peninsula, full of good habitat, might become a vital refuge for the species. Those measures might even validate a method — defense by tourniquet — that could be used on some of Tasmania's other peninsular arms.

Jones, who is a brisk, cordial woman with a mane of brown hair, picked me up in an official state Land Cruiser, and as she drove she described the effects seen so far. Her field people had culled more than a hundred devils within the past four months, she said, and though the size of the Forestier population seemed to be holding steady, the demographics had changed. Mature adults, the four- and five-year-olds, were being lost, so three-year-olds, adolescents, were accounting for most of the parenthood. The biting associated with breeding brings fatal disease, and the disease kills fast — sex equals death, a bad equation for any species. "We think extinction is a possibility within twenty-five years," Jones said.

We crossed the little bridge onto the peninsula and, after a short drive through rolling hills of eucalyptus forest, rendezvoused with the trapping crew. The chief trapper was a young woman named Chrissy Pukk, Estonian by descent, Aussie by manner, wearing a pair of blue coveralls, a dangling surgical mask, and a leather bush hat. She had been trapping devils here for three years. Jones and I tagged along as Pukk and two volunteers worked a line of forty traps placed throughout the forest. The catch rate was high, and most of the captured devils had been caught previously and injected with small electronic inserts for identification. These devils came in on a regular basis, as if the traps were soup kitchens, and Pukk recognized many of them on sight. She and only she handled the animals, cooing to them calmingly while she took their measurements, checked their body condition, and, most crucially, examined their faces for injuries and signs of tumor. One devil, a robust male Pukk called Captain Bligh, showed wounds from a recent mating session: broken teeth, a torn nose, a half-healed cut below his jaw, and a suppurating pink hole on the top of his snout, deep enough that it might have been made by a melon scoop. But he seemed to be clean of tumor. "He's just a brawler," Pukk said. She released him, and he skittered off into the brush.

"You see a lot of old friends come and go," Pukk told me as she examined another animal. For instance, there was one she had caught the day before, a male called Noddy. She had last trapped him less than a week earlier, noted inflamed whisker roots, and released him; but in the brief passage of days, those inflammations had become tumors, and now Noddy was awaiting his fate in a holding trap.

Colette Harmsen, a veterinarian who had made the long drive

south from Tasmania's Animal Health Laboratory at Mount Pleasant, was there to euthanize and autopsy any animal Pukk found unfit for release. She wore her black hair cut short, her jeans torn at the knee, a lacy black dress over the jeans, a black T-shirt reading SAVE TASSIE'S FORESTS over the dress, and, over it all, a pale blue disposable surgical smock. She was waiting at her pickup truck along with her pit bull, Lily, and her pet rat, CC, when Pukk arrived to deliver the unfortunate Noddy. Pukk and her crew returned to the trap line, Menna Jones went back to Hobart, and I stayed to watch Harmsen work. I had never seen anyone cut open a Tasmanian devil.

Her working slab was the tailgate of her pickup, spread with a clean burlap sack; her scalpels, syringes, and other tools came from a portable kit. First she anesthetized Noddy with gas. Then, after drawing blood samples from deep in his heart, she injected him with something called Lethabarb, which killed him. She measured his carcass, inspected his face, and then sliced an olive-size lump off his right cheek just below the eye. She showed me the lump's interior: a pea-size core of pale tissue surrounded by normal pink flesh. She put a chunk of it into a vial; that would go to a lab up at Mount Pleasant, she said, to be grown for chromosome typing. From the left side of the face, among the whiskers, Harmsen cut another tumor. Noddy lay limp on the burlap, both cheeks sliced away, like a halibut. When she slit open his belly and found an abundance of healthy yellow fat, she sighed. "He was in good condition." The disease hadn't progressed far. There was no sign of metastasis. But the protocols of the trapping program on Forestier don't include therapeutic surgery and chemo. Harmsen put a bit of Noddy's liver, a bit of his spleen, and a bit of his kidney into formalin. Those samples, too, would go back to Mount Pleasant for analysis. She wrapped the rest of Noddy in his burlap shroud, put him in a plastic garbage bag, and sealed that with tape. He would be incinerated. Then she cleaned up, fastidiously, to eliminate the chance that tumor cells might pass from her tailgate or her tools to another animal.

The phenomenon of transmissible tumors isn't confined to canines, Tasmanian devils, and Syrian hamsters. There have been human cases, too. Forty years ago a team of physicians led by Edward F. Scanlon reported, in the journal *Cancer*, that they had "decided to transplant small pieces of tumor from a cancer patient into a healthy donor, on a well informed volunteer basis, in the hope of gaining a little better understanding of cancer immunity," which they thought might help in treating the patient. The patient was a fifty-year-old woman with advanced melanoma; the "donor" was her healthy eighty-year-old mother, who had agreed to receive a bit of the tumor by surgical transplant. One day after the transplant procedure, the daughter died suddenly from a perforated bowel. Scanlon's report neglects to explain why the experiment wasn't promptly terminated — why they didn't dive back in surgically to undo what had been done to the mother. Instead, three weeks were allowed to pass, at which point the mother had developed a tumor indistinguishable from her daughter's. Now it was too late for surgery. This cancer moved fast. It metastasized, and the mother died about fifteen months later, with tumors in her lungs, ribs, lymph nodes, and diaphragm.

The case of the daughter–mother transplant and the case of the Syrian hamsters have one common element: the original sources of the tumor and the recipients were genetically very similar. If the genome of one individual closely resembles the genome of another (as children resemble their parents and as inbred animals resemble one another), the immune system of a recipient may not detect the foreignness of transplanted cells. The hamsters were highly inbred (intentionally, for experimental control) and therefore not very individuated from one another as far as their immune systems could discern. The mother and daughter were also genetically similar — as similar as two people can be without being identical twins. Lack of normal immune response, because of such closeness, goes some way toward explaining why those tumors survived transference between individuals.

Low immune response also figures in two other situations in which tumor transmission is known to occur: pregnancy and organ transplant. A mother sometimes passes cancer cells to her fetus in the womb. And a transplanted organ sometimes carries tiny tumors into the recipient, vitiating the benefits of receiving a lifesaving liver or kidney from someone else. Cases of both kinds are very rare, and they involve some inherent or arranged compatibility between the original victim of the tumor and the secondary victim, plus an immune system that is either compromised (by immunosuppressive drugs in the organ recipient) or immature (in the fetus).

Other cases are less easily explained. In 1986 two researchers from the National Institutes of Health reported that a laboratory worker, a healthy nineteen-year-old woman, had accidentally jabbed herself with a syringe carrying colon-cancer cells; a colonic tumor grew in her hand, but she was rescued by surgery. More recently, a fifty-three-year-old surgeon cut his left palm while removing a malignancy from a patient's abdomen, and five months later he found himself with a palm tumor, one that genetically matched the patient's tumor. His immune system responded, creating an inflammation around the tumor, but the response was insufficient and the tumor kept growing. Why? How? It wasn't supposed to be able to do that. Again, though, surgery delivered a full cure. And then there's Henri Vadon. He was a medical student in the 1920s who poked his left hand with a syringe after drawing liquid from the mastectomy wound of a woman being treated for breast cancer. Vadon, too, developed a hand tumor. Three years later, he died of metastasized cancer because neither the surgical techniques of his era nor his own immune system could save him.

The tumor that I had watched Colette Harmsen harvest from Noddy's face would be examined at the Mount Pleasant labs by Anne-Maree Pearse and her assistant Kate Swift. Pearse is a former parasitologist, now working in cell biology, and she has a special interest in the genetics of devil facial tumor disease. She and Swift were the researchers who in 2006 published a dramatic report in the British journal *Nature* that, with eight paragraphs of text and a single photographic image, had answered the lingering question about whether DFTD is a genuinely transmissible cancer.

Pearse came out of retirement (she had turned to running a flower farm) in response to the scientific conundrum of DFTD. A back injury has forced her to use of a cane, but she is vigorous when describing her research. Although she was originally trained as an entomologist, her work with fleas drew her into the world of parasitology, and from there it was just a few more steps into oncology and the study of lymphomas among the devil and its close marsupial relatives. "Somehow my whole life was preparing me for this," she said when I visited her lab at Mount Pleasant. She added, almost appreciatively, "This disease." Pearse tends to think, as she put it, "outside the square" — a useful trait in the case of DFTD, she said, because the disease isn't behaving like anything hereto-fore known. "It's a parasitic cancer," she told me. "The devil's the host."

For the 2006 study, Pearse and Swift examined chromosome structure in tumor cells from eleven different devils. They found that the tumor chromosomes were abnormal (misshapen, some missing, some added) compared with those from healthy devil cells, but that the tumor chromosomes, from one cell or another, from one tumor or another, were abnormal in *all the same ways.* You could see that comparison graphically in the photo in *Nature:* fourteen nice sausages matched against thirteen variously mangled ones. Those thirteen chromosomes, wrote Pearse and Swift, had undergone "a complex rearrangement that is identical for every animal studied." The mangling was unmistakable evidence. It ap-peared in each tumor but not elsewhere in each animal. "In light of this remarkable finding and of the known fighting behavior of the devils," Pearse and Swift wrote, "we propose that the disease is transmitted by allograft" — tissue transplant — "whereby an infec-tious cell line is passed directly between the animals through bites they inflict on one another."

Pearse and Swift had proved that DFTD is a highly infectious form of cancer, its transmission made possible by, among other fac-tors, the habit of mutual face-biting. When I visited Menna Jones, she expressed the same idea: "It's a piece of devil tissue that be-haves like a parasite." Jones was using the word "parasite" in its strict biological sense, meaning any organism that lives on or within another kind of organism, extracting benefit for itself and causing harm to the other. The first rule of a successful parasite is, don't kill your host — or, at least, don't kill one host until you've had time to leap aboard another. DFTD, passing quickly from devil to devil, killing them all but not quite so quickly, follows that rule.

How does any parasite, whether it is a species or merely a tumor, ac-quire the attributes and tactics necessary for survival, reproduc-tion, and continuing success? The answer is simple but not obvious: evolution.

Cancer and evolution have traditionally been considered sepa-rately by different scientists with different interests using different methods. You could graduate from medical school, you could follow that with a Ph.D. in cell biology or molecular genetics, you could become a respected oncologist or a well-funded cancer researcher, without ever having read Darwin. You could do it, in fact, without having studied much evolutionary biology at all. Many cell and molecular biologists tended even to scorn evolutionary biology as a "merely descriptive" enterprise, lacking the rigor, quantifiability, and explanatory power of their disciplines. There were exceptions to this disconnect, cancer scientists who even during the early days thought in evolutionary terms, but those scientists had little influence.

In recent decades, however, the situation has changed, as molecular genetics and evolutionary biology have converged on some shared questions. One signal act of synthesis occurred in 1976 when a leukemia researcher named Peter Nowell published a theoretical paper in *Science* titled "The Clonal Evolution of Tumor Cell Populations." Nowell proposed what was then a novel idea: that the biological events occurring when cells progress from normal to precancerous to cancerous represent a form of evolution by natural selection. As with the evolution of species, he suggested, the evolution of malignant tumors requires two conditions: genetic diversity among the individuals of a population and competition among those individuals for limited resources. Genetic diversity within one mass of precancerous cells comes from mutations — copying errors and other forms of change — that yield variants as the cells reproduce. That is, in the very act of replicating themselves (sometimes inaccurately), the cells diversify into a population encompassing some small genetic differences between one cell and another. Each variant cell then replicates itself true to type, constituting a clonal lineage (a lineage of accurate copies), until the next mutation creates a new variant. The fittest variants survive and proliferate. By this means, the genetic character of the cell population gradually changes, and with such change comes adaptation, a better fit to environmental circumstances. What constitutes "the fittest" among clonal lineages within a precancerous growth? Those that can reproduce fastest. Those that can resist chemotherapy. Those that can metastasize and therefore escape the surgeon's knife.

Nowell's hypothesis about tumor evolution became widely known and accepted within certain circles of cancer research. (Among

other researchers, it wasn't adamantly disputed but merely ignored.) Those circles have more recently produced a lot of rich
theorizing and a smaller amount of empirical work, supporting
Nowell and carrying his idea forward. A culmination of sorts occurred in 2000, when the cancer geneticist Robert Weinberg, discoverer of the first human oncogene and the first tumor suppressor gene, published a concise paper titled "The Hallmarks of
Cancer." Weinberg and his coauthor, Douglas Hanahan, described
six "acquired capabilities," such as endless self-replication, the ignoring of antigrowth signals, the invasion of neighboring tissues,
and the refusal to die, that collectively characterize cancer cells.
How are those capabilities acquired? By mutations and other genetic changes, giving cells with one such trait or another competitive advantage over normal cells. Hanahan and Weinberg added
that "tumor development proceeds via a process formally analogous to Darwinian evolution." With this cautious phrasing, they
gave authoritative endorsement to the idea that Peter Nowell had
proposed: cancers, like species, evolve.

In 1998 a young researcher named Carlo Maley began looking for a
way to study the evolution of cancer. Educated at Oxford and MIT
as an evolutionary biologist and a computer modeler, Maley had
no training in medicine and not much in molecular biology. During a postgraduate fellowship, though, he became interested in infectious disease. He figured that if evolution was cool, then coevolution — wherein both parasite and host are evolving — would
be doubly cool. Then he stumbled across a description of cancer
as an evolving disease. He read that Sir Walter Bodmer, a British
geneticist and the former director of the Imperial Cancer Research Fund, had urged his cancer-research colleagues to "think
evolution, evolution, evolution" when they considered tumor cells.
Maley typed "evolution and cancer" into a search engine for the
scientific literature, which turned up very little. He did learn of
Nowell's hypothesis, but that was just theory. He was groping. He
had done plenty of theoretical modeling, but for this task he
needed the desperate realities, and the data, of clinical oncology.
    And then, at a workshop in Seattle, Maley met Brian Reid, an experienced cancer researcher studying something called Barrett's
esophagus, a precancerous condition of the lower throat. They hit
it off. Reid had the right clinical situation but wasn't deeply versed

in evolutionary biology; Maley had the right background. They agreed to collaborate.

Reid and his colleagues possessed sixteen years of continuous data on Barrett's patients and a tissue bank going back to 1989. They knew which patients had developed esophageal cancer and which hadn't, and they could match those outcomes against what they had seen in cell cultures and genetic work from earlier in the patients' history. So they could ask evolutionary questions that were answerable from patterns in the data. The most basic question was: did tumors become malignant through evolution by natural selection? The other big question was: can doctors predict which precancerous growths will turn malignant? Maley and Reid, along with additional collaborators, found that case histories of Barrett's esophagus tend to confirm Nowell's hypothesis. Cancerous tumors, like species, *do* evolve. And from the Barrett's data, predictions can be made. The higher the diversity of different cell variants within a precancerous growth, the greater the likelihood that the growth will progress to malignancy. Why? Because of the basic Darwinian mechanism. Genetic diversity plus competitive struggle eliminates unfit individuals and leaves the well-adapted to reproduce.

Maley and Reid have more recently taken such thinking one step beyond evolution — into ecology. Along with Lauren M. F. Merlo (as first author) and John W. Pepper, they published a provocative paper titled "Cancer as an Evolutionary and Ecological Process," in which they discussed not just tumor evolution but also the ecological factors that form evolution's context, such as predation, parasitism, competition, dispersal, and colonization. Dispersal is travel by venturesome individuals, which in some cases allows species to colonize new habitats. Merlo, Maley, and their colleagues noted three ways in which the concept of dispersal is applicable to cancer: small-scale cell movement within a tumor (not very important), invasion of neighboring tissues (important), and metastasis (fateful).

Reading that, I remembered devil facial tumor disease and wondered whether there might not be a fourth way: transmissibility. An infectious cancer is a successful disperser. It colonizes new habitat. DFTD seems to be dispersing and colonizing, much as pigeons disperse across oceans, colonizing new islands. This wasn't just evolution; it was evolutionary ecology.

I called one of the paper's coauthors, John W. Pepper, an evolutionary biologist at the University of Arizona, and asked whether I was stretching the notion too far. No, he said, you're not. If he could revise that paper again, Pepper told me, he would insert the idea that tumors evolve toward transmissibility.

Eight hundred million years ago there was no such thing as cancer. Virtually all living creatures were single-cell organisms, and the rule was *Every cell for itself!* Uncontrolled, undifferentiated cell growth wasn't abnormal. It was the program of all life on Earth.

Then, around 700 million years ago, things changed. Paleontologists call this event the Cambrian explosion. Complex multicellular animals, metazoans, appeared. And not just meta*zoans* but meta-*phytes,* too — that is, multicellular plants. How did it happen? Very gradually, as single-cell creatures resembling bacteria or algae began to aggregate into colonial units and discover, by trial and error, how they could benefit from division of labor and specialization of shape and function. To enjoy those benefits, they had to set aside the old rule of absolute selfishness. They had to cooperate. They couldn't cheat against the interests of the collective entity. (Or anyway they *shouldn't* cheat very often; otherwise the benefits of collectivity wouldn't accrue.) Cooperation was a winning formula. Primitive multicellular creatures, roughly along the lines of jellyfish or sponges or slime molds, began to succeed, to grow, to occupy space, and to claim resources in ways that loner cells couldn't. You can see their imprints in the Burgess Shale: weird things like sci-fi vermin, prevertebrate, preinsect, that seem to have been built out of bubble wrap and old Slinkys. They succeeded for a while, then gave way to still better designs. Multicellularity offered wide possibilities.

But uncontrolled cell replication didn't disappear entirely. Sometimes a single atavistic cell would ignore the collective imperative; it would revert to the old habit — proliferating wildly, disregarding all signals to stop. It would swell into a big, greedy lump of its own kind, and in so doing disrupt one or more of the necessary collective functions. That was cancer.

The risk of runaway cell replication remained a factor in the evolutionary process, even as multicellular creatures increased vastly in complexity, diversity, size, and dominance on our planet. And species responded to that risk just as they responded, incrementally and over long periods of time, to other risks, such as predation or parasitism: by acquiring defenses. One such defense is the amazing ability of living cells to repair mutated DNA, putting the cell program back together properly after a mishap during cell replication. Another defense is *apoptosis,* a form of programming that tells a cell not to live forever. Another is cellular senescence, during which a cell continues to live but is no longer capable of replicating. Another is the distinction between stem cells and differentiated cells, which limits the number of cells responsible for cell-replacement activity and thereby reduces the risk of accumulated mutations. Another is the requirement for biochemical growth signals before a cell can begin to proliferate. Many of these defenses are controlled by tumor suppressor genes, such as the one that produces a protein that prevents cells from replicating damaged DNA. Nobody knows just how many anticancer defenses exist within a given species (we humans seem to have more than mice do, and possibly not so many as whales), but we do know that they make our continuing lives possible.

Tumors, in the course of their own evolution from one normal cell to a cancerous malignancy, circumvent these natural defenses. They may also change in response to externally imposed defenses, such as surgery, chemotherapy, and radiation. The fittest cells, in Darwinian terms, are those that reproduce themselves most quickly and aggressively, resisting all signals to desist and all attempts to kill them. The victim (that is, the human or the Tasmanian devil or the Syrian hamster) suffers the consequences, having become the arena for an evolutionary struggle at a scale far different from that of its own struggle to survive. But the principles of the struggle are the same at each scale.

This process, whereby cells mutate, reproduce, and proliferate differentially within a body, is called somatic evolution. It stands as a counterpoint to organismic evolution (progressive changes at the scale of whole bodies within a population), and the opportunities for it to proceed are abundant. According to one count, at least 291 genes in the human body contribute, when damaged by mutation, toward somatic evolution.

Mutations occur when something goes wrong during cell replication. A cell replicates by copying its DNA (sometimes inaccurately), sorting the DNA into two identical parcels of chromosomes, then splitting into two new cells, each with its own chromosomes. This process is called mitosis. The goal is not to generate an ever-higher number of cells during a creature's adult lifetime but simply to re-place old cells with new ones. Mitosis counterbalanced with apop-tosis, cell death, should provide a constant supply. But each time a mitotic division occurs, there is some very small chance of muta-tion. And the many small chances add up. A human body contains about 30 trillion cells. The number of cell divisions that occur in a lifetime is far larger: 10,000 trillion. A disproportionately large share of those divisions occurs in epithelial cells, which serve as boundary layers or linings, such as the skin and the interior surface of the colon. That's why skin cancer and colon cancer are relatively common — more cell turnover, more chance of mutation and evo-lution.

How many mutations does it take for a malignancy to occur? Esti-mates range from three to twelve in humans, depending on the form of cancer. Five or six is considered an operational average for purposes of discussion. Here's some good news: for a cell to ac-quire those five or six changes, at the usual rate of mutation, is highly, highly, highly unlikely. The odds are great against quintuple mutation in any given cell, making cancer seem impossible within a human life span. One form of mutation, however, can vastly in-crease the later rate of mutation, which gives the precancerous cells many more chances to become malignant.

In the United States, about 40 percent of us will eventually get cancer bad enough to be diagnosed. And autopsies suggest that vir-tually all of us will be nurturing incipient thyroid cancer by the time we die. Among octogenarian and nonagenarian men, 80 per-cent carry prostate cancer when they go. Cancer is terrible, cancer is dramatic, but cancer isn't rare. In fact, it's nearly universal.

The biological mystery of how the Tasmanian devil's rogue tumor manages to establish itself in one animal after another is still un-solved. But a good hypothesis has been offered by an immunologist named Greg Woods at the University of Tasmania. Woods and his group studied immune reactions in *Sarcophilus harrisii*, which seem generally to be normal against ordinary sorts of infection. Against DFTD cells, though, no such reaction occurs. "The tumor is just not *seen* by the immune system, because it just looks too similar," Woods told me when I stopped by his lab. The devils have low ge-netic diversity, probably because they inhabit a small island, they colonized it originally by way of just a few founders, and they have passed through some tight population bottlenecks in the centuries since. They're not quite so alike one another as a bunch of inbred hamsters, but they're too alike for their own good in the current sad, anomalous circumstances. Their immune systems don't reject the tumor cells because, Woods suspects, in each animal the critical MHC genes (the major histocompatibility complex, which produces proteins crucial to immunological policing) are all virtually identical, and the devils' police cells can't distinguish "him" from "me."

Most of the DFTD team are, like Greg Woods and Anne-Maree Pearse, located in Hobart or Launceston. Most of the animals aren't. So, two days after the autopsy on Noddy, I drove back to the Forestier Peninsula for another round of trapping with Chrissy Pukk and her crew. It wasn't that I expected to learn any new angles on the science. I just wanted to see more Tasmanian devils.

This time, Pukk issued me my own pair of blue coveralls. As we set off along the trap line, she exuded the contentment of a joyously crude tomboy enjoying the best job in the world: trapping devils for the good of the species. The only downside was the necessity of issuing a death sentence to any animal with a trace of tumor. "You get attached to the individuals," she admitted. "But you've got to remember all the other individuals you can save if you take that animal out early on."

Wouldn't this be less difficult emotionally, I asked her, if you gave them *numbers* instead of names? She answered the question — saying she couldn't do her work properly if she wasn't emotionally invested, plus which, names were easier to remember — and then she continued to answer it throughout the day. These creatures, they all have their memorable eccentricities, their little histories, she explained. Some she could recognize almost by smell. You couldn't do *that* with numbers.

There were forty traps again today, and about a dozen trapped devils to process, all recaptures, previously tagged and named. Trap by trap, animal by animal, Pukk worked through the measurements and the facial exams, handling each devil firmly but with a steady touch that provoked no devilish squirming: Captain Bligh (looking glum, or maybe a little embarrassed at having been caught again so soon), Hipster, Isabel, Masikus (Estonian for "strawberry"), Miss Buzzy Bum (her rump had been full of burrs), Rudolph (thus called for a nose that had been rubbed raw), Sandman, Skipper, and many others. They may have been virtually indistinguishable in the terms by which immune systems operate, but Chrissy Pukk knew each devil at a glance.

Rudolph's condition gave her pause. He was a two-year-old, nicely grown since she had first trapped him, his red nose healed . . . but there was something on the edge of his right eye. A pink growth, no bigger than a caper. "Oh shit," she said. Tumor? Or maybe it was just a little wound, puffy and raw. She looked closely. She peered into his mouth. She palpated lymph nodes at the base of his jaw. The volunteers and I waited in silence. Evolution had shaped Rudolph for survival, and evolution might take him away. It was all evolution: the yin of struggle and death, the yang of adaptation, DFTD versus *Sarcophilus harrisii*. The leaping tumor, well adapted for fast replication and transmissibility, has its own formidable impulse to survive. And no one could know at this point, not even Chrissy, whether it had already leapt into Rudolph.

"Okay," she said, sounding almost sure of her judgment, "I'm gonna give him the all clear." She released him and he ran.

At latest report, devil facial tumor disease has spread across 60 percent of Tasmania's land surface, and in some areas, especially where it got its earliest start, the devil population seems to have declined by as much as 90 percent. In November 2007 the Tasmanian government classified the devil as "endangered." DFTD specialists differ strongly on how such a crisis should be met. One view is that suppressing the disease — trapping and euthanizing as many infected animals as possible and then establishing barriers, as on the Forestier Peninsula — is the best strategy. Another view is that the species, virtually doomed on mainland Tasmania, can better be saved by transplanting disease-free devils to a small offshore island. Still another view, maybe the boldest and most risky, is that doing nothing — allowing the disease to spread unchecked — might yield a small remnant population of survivors with natural immunity to DFTD who could repopulate Tasmania.

Weeks after my last outing with the trappers, back in the Northern Hemisphere and wanting a broader perspective — not just on the fate of the devils but on the evolution of cancer — I met Robert Weinberg at a stem-cell conference in Big Sky, Montana. Because it was a Sunday, with the first session not yet convened, and because he is a genial man, Weinberg gave me a two-hour tutorial in a boardroom of the ski lodge, fortifying some points by flipping through his own four-pound textbook, *The Biology of Cancer,* a copy of which I had lugged to our meeting like a student. He was incognito in a plaid Woolrich shirt. He'd be called on that evening to deliver the keynote address, but never mind, he was prepared.

"Infectious cancer is really an aberration," Weinberg told me, affirming what Greg Woods had said. "It's so bizarre. It has happened only rarely." Maybe it's possible only in cases where there's close physical contact between susceptible tissues. "That, right away, limits it to venereal tumors or tumors that can be transmitted by biting." Weinberg knew that I'd walked in with a head full of Tasmanian devils.

Does this mean that cancer cells are harder to transmit than, say, virus particles? "Much," he said. "Cells are very *effete*. Very susceptible to dying in the outside world." They dry out, they wither, they don't remain viable when they're naked and alone. Bacteria can form spores. Viruses in their capsules can lie dormant. But cells from a metazoan? No. They're not packaged for transit.

And that's only one of two major constraints, Weinberg said. The second is that if cancer cells *do* pass from one body to another, they are instantly recognized as foreign and eliminated by the immune system. Each cell of any sort bears on its exterior a set of protuberant proteins that declare its identity; they might be thought of as its travel papers. These proteins are called antigens and are produced uniquely in each individual by the MHC (major histocompatibility complex) genes. If the travel papers of a cell are unacceptable (because the cell is an invader from some other body), the T cells (one type of immunological police cell) will attack and obliterate it. If the invader cell shows no papers at all, another kind of police cell (called NK cells) will bust it. Only if the antigens on the cell surface have been "down-regulated" discreetly but not eliminated altogether can a foreign cell elude the immune system of a host. That's what canine transmissible venereal tumor seems to have done: down-regulated its antigens. It shows fake travel papers — blurry, faded, but just good enough to get by.

Nice trick! How did CTVT do that? Although nobody knows exactly, the best hypothesis is evolution by natural selection — or by some process "formally analogous" to it.

Weinberg went on to explain that the process is a little more complicated than classic Darwinian selection. Darwin's version works by selection among genetic variations that differentiate one organism from another, and in sexually reproducing species those variations are heritable. But evolution in tumor lineages occurs by that sort of selection plus another sort — selection among *epigenetic* modifications of DNA. Epigenetic means outside the line of genetic inheritance: acquired by experience, by accident, by circumstance. Such secondary chemical changes to the molecule affect behavior, affect shape, and pass from one cell to another but do *not*, contrary to the analogy, pass from parent to offspring in sexual reproduction. These changes are peeled away in the process of meiosis (the formation of sperm and egg cells for sexual reproduction) but preserved in mitosis (the process of simple cell replication in the body). So cancerous cell reproduction brings such changes forward into the new cells, along with the fundamental genetic changes.

Does that mean tumors don't evolve? Certainly not. They do. "It's still Darwin," Weinberg said. "It's Darwin revised."

JOSHUA ROEBKE

The Reality Tests

FROM *Seed*

To ENTER THE SOMEWHAT FORMIDABLE Neo-Renaissance
building at Boltzmanngasse 3 in Vienna, you must pass through a
small door sawed from the original cathedral-like entrance. When I
first visited this past March, it was chilly and overcast in the late afternoon. Atop several tall stories of scaffolding there were two men
who would hardly have been visible from the street were it not for
their sunrise-orange jumpsuits. As I was about to pass through the
nested entrance, I heard a sudden rush of wind and felt a mist of
winter drizzle. I glanced up. The veiled workers were power-wash-
ing away the building's façade, down to the century old brick underneath.

In 1908 Karl Kupelwieser, Ludwig Wittgenstein's uncle, donated
the money to construct this building and turn Austria-Hungary
into the principal destination for the study of radium. Above the
doorway the edifice still bears the name of this founding purpose.
But since 2005 this has been the home of the Institut für Quan-
tenoptik und Quanteninformation (IQOQI, pronounced "ee-ko-
kee"), a center devoted to the foundations of quantum mechanics.
The IQOQI, which includes a sister facility to the southwest in the
valley town of Innsbruck, was initially realized in 2003 at the behest
of the Austrian Academy of Sciences. However, the institute's conception several years earlier was predominantly due to one man:
Anton Zeilinger.

In January 2008, Zeilinger became the first-ever recipient of the
Isaac Newton Medal for his pioneering contributions to physics as
the head of one of the most successful quantum optics groups in

the world. Over the past two decades, he and his colleagues have done as much as anyone else to test quantum mechanics. And since its inception more than eighty years ago, quantum mechanics has possibly weathered more scrutiny than any theory ever devised. Quantum mechanics appears correct, and now Zeilinger and his group have started experimenting with what the theory means.

Some physicists still find quantum mechanics unpalatable, if not unbelievable, because of what it implies about the world beyond our senses. The theory's mathematics is simple enough to be taught to undergraduates, but the physical implications of that mathematics give rise to deep philosophical questions that remain unresolved. Quantum mechanics fundamentally concerns the way in which we observers connect to the universe we observe. The theory implies that when we measure particles and atoms, at least one of two long-held physical principles is untenable: distant events do not affect one other, and properties we wish to observe exist before our measurements. One of these, locality or realism, must be fundamentally incorrect.

For more than seventy years, innumerable physicists have tried to disentangle the meaning of quantum mechanics through debate. Now Zeilinger and his collaborators have performed a series of experiments that, while neatly agreeing with the theory's predictions, are reinvigorating these historical dialogues. In Vienna experiments are testing whether quantum mechanics permits a fundamental physical reality. A new way of understanding an already powerful theory is beginning to take shape, one that could change the way we understand the world around us. Do we create what we observe through the act of our observations?

Most of us would agree that there exists a world outside our minds. At the classical level of our perceptions, this belief is almost certainly correct. If your couch is blue, you will observe it as such whether drunk, in high spirits, or depressed; the color is surely independent of the majority of your mental states. If you discovered that your couch was suddenly red, you could be sure there was a cause. The classical world is real, and not only in your head. Solipsism hasn't really been a viable philosophical doctrine for decades, if not centuries.

But none of us perceives the world as it exists *fundamentally*. We do not observe the tiniest bits of matter, nor the forces that move them, individually through our senses. We evolved to experience the world in bulk, our faculties registering the net effect of trillions upon trillions of particles or atoms moving in concert. We are crude measurers. So divorced are we from the activity beneath our experience that physicists became relatively assured of the existence of atoms only about a century ago.

Physicists attribute a fundamental reality to what they do not directly perceive. Particles and atoms have observable effects that are well described by theories like quantum mechanics. Single atoms have been "seen" in measurements and presumably exist whether or not we observe them individually. The properties that define particles — mass, spin, and so on — are also thought to exist before we measure them. In physics this is how reality is defined; particles and atoms have measurable properties that exist prior to measurement. This is nothing stranger than your blue couch.

As a physical example, light consists of particles known as photons that each have a property called polarization. Measuring polarization is usually something like telling time; the property can be thought of like the direction of a second hand on a clock. For unpolarized light, the second hand can face any direction, as on a normal clock; for polarized light, the hand will face in only one or a few directions, as if the clock were broken. That photons can be polarized is, in fact, what allows some sunglasses to eliminate glare — the glasses block certain polarizations and let others through. In Vienna the polarization of light is also being used to test reality.

For a few months in 2006, Simon Gröblacher, who had started his Ph.D. not long before, spent his Saturdays testing realism. Time in the labs at the IQOQI is precious, and during the week other experiments with priority were already underway. Zeilinger and the rest of their collaborators weren't too worried that this kind of experiment would get scooped. They were content to let Gröblacher test reality in the lab's spare time.

It was after two P.M. when I first met Gröblacher, and he had just woken up; they are installing an elevator in his lab, so he works nights. He had told me to come to the top floor of the IQOQI building to find him. I made my way up the broad granite steps, and on the final landing I heard shouts from a half-open door. There was a raucous game of foosball in the lounge. When Gröblacher saw me, someone else grabbed the handles.

The lab where Gröblacher performed the first experiment on realism is on the second floor of the Universität Wien physics department, which connects to the IQOQI through a third-floor bridge. The original experiment has given way to another, but, Gröblacher tells me, the setup looks roughly the same.

In the middle of the cramped space is a floating metal surface, about the size of a banquet table, latticed with drill holes. A forest of black optical equipment, like monocles atop tiny poles, seems to grow out of the table. Beam splitters resemble exact, glass dies. In the center is an encased crystal that is not visible, and on the ends sit idle lasers.

Gröblacher walked me through the tabletop obstacle course: the laser light passes through a series of polarizers and filters, hits the crystal, and splits into two beams of single-file photons. Detectors in both beams measure the polarization of each photon, which are related to one another. The data are tested against two theories: one that preserves realism but allows strange effects from anywhere out there in the universe, and quantum mechanics.

The whole experiment would fit snugly in a child's bedroom, and as I looked at the table, I refrained from asking my first instinctual questions. "This is it? This is where you tested realism?" I already knew how unfair these questions were. It had taken a few months of tests and almost two years for Zeilinger's group to understand how this experiment tests realism. Before that, it had been more than eighty years since physicists began to argue about what quantum mechanics had to do with reality at all.

In the summer of 1925, Werner Heisenberg was stricken with hay fever and having trouble with math. He asked his adviser for two weeks off and left for a barren island in the North Sea. He spent his mornings swimming and hiking, but every evening Heisenberg tried to describe atoms in a theory that included only what could be measured. One night, feverish with insight, he calculated until dawn. After Heisenberg put down his pencil as the sun began to rise, he walked to the tip of the island, confident he had discovered quantum mechanics.

By this time a quarter-century had passed since Max Planck first described energy as whole-number multiples of a basic unit, which he called the quantum. When two of the quantum's other leading progenitors, Niels Bohr and Albert Einstein, heard about Heisenberg's completion of the work they began, their reactions were almost immediate; Bohr was impressed, Einstein was not. Heisenberg's theory emphasized the discrete, particle-like nature of matter, and Einstein, who tended to think in images, could not picture it in his head.

In Switzerland, Erwin Schrödinger had also been "repelled" by Heisenberg's theory. In the fall of 1925, Schrödinger was thirty-eight years old and rife with self-doubt, but when Einstein sent him an article describing a possible duality between particles and waves, Schrödinger had an idea. Over a period of six months, he published five papers outlining a wave theory of the atom. Though it proved difficult to physically interpret what his wave was, the theory felt familiar to Schrödinger. Heisenberg, who had moved to Copenhagen to become Bohr's assistant, thought the theory "disgusting."

Schrödinger and Heisenberg independently uncovered dual descriptions of particles and atoms. Later the theories proved equivalent. Then in 1926 Heisenberg's previous adviser, Max Born, discovered why no one had found a physical interpretation for Schrödinger's wave function. They are not physical waves at all; rather, the wave function includes all the possible states of a system. Before a measurement those states exist in *superposition*, wherein every possible outcome is described at the same time. Superposition is one of the defining qualities of quantum mechanics and implies that individual events cannot be predicted; only the probability of an experimental outcome can be derived.

The following year, 1927, Heisenberg discovered the uncertainty principle, which placed a fundamental limit on certain measurements. Pairs of specific quantities are incompatible observables; momentum and position, energy and time, and other measurable pairs cannot be known together with absolute accuracy. Measuring one restricts knowledge of the other. With this, quantum mechanics had become a full theory. But what physicists ended up with was a world divided. There was an inherent distinction between atoms unseen and their collective motion, which we witness with our eyes — the quantum versus the classical. While the distinction appeared physical, many, like Bohr, thought it philosophical; the theory lacked a proper interpretation.

According to Bohr, every measuring device affects what it is used to observe. The quantum world is discrete, so there can never be absolute precision during a measurement. To know about quantum mechanics, we rely on classical devices. To Bohr this implied that the hierarchy between observer and observed had no meaning; they were nonseparable. Concepts once thought to be mutually exclusive, such as waves and particles, were also complements. The difference was only language.

By contrast, Einstein was a realist who believed in a world independent of the way it is measured. During a set of conferences at the Hotel Metropole in Brussels, he and Bohr argued famously over the validity of quantum mechanics, and Einstein presented a number of thought experiments intended to show the theory incorrect. But when Bohr used Einstein's own theory of relativity to evade one of these thought experiments, Einstein was so stung he never tried to disprove quantum mechanics again, though he continued to criticize it.

In 1935, from an idyllic corner of New Jersey, Einstein and two young collaborators began a different assault on quantum mechanics. Einstein, Podolsky, and Rosen (EPR) did not question the theory's correctness, but rather its completeness. More than the notion that God might play dice, what most bothered Einstein were quantum mechanics' implications for reality. As Einstein prosaically inquired once of a walking companion, "Do you really believe that the moon exists only when you look at it?"

The EPR paper begins by asserting that there's a real world outside theories. "Any serious consideration of a physical theory must take into account the distinction between the objective reality, which is independent of any theory, and the physical concepts with which the theory operates." If quantum mechanics is complete, then "every element of physical reality must have a counterpart in the physical theory." EPR argued that objects must have preexisting values for measurable quantities and that this implied that certain elements of reality could not be determined by quantum mechanics.

Einstein and his colleagues imagined two electrons that collide and fly apart. After the collision the electrons exist in a state of superposition of the possible values for their momenta. Mathematically and physically, it makes no sense to say that either electron has a definite momentum independent of the other before measurement; they are "entangled." But when one electron's momentum is measured, the value of the other's is instantly known and the superpositions collapse. Once the momentum is known for a particle, we cannot measure its position. This element of reality is denied us by the uncertainty principle. Even stranger is that this occurs even when the electrons fly vast distances apart before measurement. Quantum mechanics still describes the electrons as a single system across space. Einstein could never stomach that an experiment at one electron would instantaneously affect the other.

In Copenhagen Bohr began an immediate response. It didn't matter if particles might affect one another over vast distances, or that particles had no observable properties before they are observed. As Bohr later said, "There is no quantum world. There is only an abstract quantum physical description."

Physicists' discourse on reality began just as the world slid inexorably toward war. During World War II physicists once interested in philosophy worried about other issues. David Bohm, however, did worry. After the war Bohm was a professor at Princeton, where he wrote a famous textbook on quantum mechanics. Einstein thought it was the best presentation of quantum mechanics he had read, and when Bohm began to challenge the theory, Einstein said, "If anyone can do it, then it will be Bohm."

In 1952, during the Red Scare, Bohm moved to Brazil. There he discovered a theory in which a particle's position was determined by a "hidden variable" even when its momentum was absolutely known. To Bohm reality was important, and so, to preserve it, he was willing to abandon locality and accept that entangled particles influenced one another over vast distances. However, Bohm's hidden-variables theory made the same predictions as quantum mechanics, which already worked.

In America Bohm's theory was ignored. But when the Irishman John Bell read Bohm's idea, he said, "I saw the impossible done." Bell thought hidden variables might show quantum mechanics incomplete. Starting from Bohm's work, Bell derived another kind of hidden-variables theory that could make predictions different from those of quantum mechanics. The theories could be tested against one another in an EPR-type experiment. But Bell made two assumptions that quantum mechanics does not; the world is local (no distant influences) and real (preexisting properties). If quantum mechanics was correct, one or both of these assumptions were false, though Bell's theorem could not determine which.

Bell's work on local hidden-variables theory stirred little interest until the 1970s, when groups lead by John Clauser, Abner Shimony, and others devised experimental schemes in which the idea could be tested with light's polarizations instead of electrons' momentum. Then in 1982 a young Frenchman named Alain Aspect performed a rigorous test of Bell's theory on which most physicists finally agreed. Quantum mechanics was correct, and either locality or realism was fundamentally wrong.

During the 1980s and 1990s, the foundations of quantum mechanics slowly returned to vogue. The theory had been shown, with high certainty, to be true, though loopholes in experiments still left some small hope for disbelievers. However, even to believers, nagging questions remained: was the problem with quantum mechanics locality, realism, or both? Could the two be tested?

In May of 2004 Markus Aspelmeyer met Anthony Leggett during a conference at the Outing Lodge in Minnesota. Leggett, who had won the Nobel Prize the year before, approached Aspelmeyer, who had recently become a research assistant to Zeilinger, about testing an idea he first had almost thirty years before.

In 1976 Leggett left Sussex on a teaching exchange to the University of Science and Technology in Kumasi, the second largest city in Ghana. For the first time in many years, he had free time to really think, but the university's library was woefully out of date. Leggett decided to work on an idea that didn't require literature because few had thought about it since David Bohm: nonlocal hidden-variables theories. He found a result, filed the paper in a drawer, and didn't think about it again until the early 2000s.

Leggett doesn't believe quantum mechanics is correct, and there are few places for a person of such disbelief to now turn. But Leggett decided to find out what believing in quantum mechanics might require. He worked out what would happen if one took the idea of nonlocality in quantum mechanics seriously, by allowing for just about any possible outside influences on a detector set to register polarizations of light. Any unknown event might change what is measured. The only assumption Leggett made was that a natural form of realism hold true; photons should have measurable polarizations that exist before they are measured. With this he laboriously derived a new set of hidden-variables theorems and inequalities, as Bell once had. But whereas Bell's work could not distinguish between realism and locality, Leggett's did. The two could be tested.

When Aspelmeyer returned to Vienna, he grabbed the nearest theorist he could find, Tomasz Paterek, whom everyone calls "Tomek." Tomek was at the IQOQI on fellowship from his native Poland, and together they enlisted Simon Gröblacher, Aspelmeyer's student. With Leggett's assistance, the three spent six months painfully checking his calculations. They even found a small error. Then they set about recasting the idea, with a few of the other resident theorists, into a form they could test. When they were done, they went to visit Anton Zeilinger. The experiment wouldn't be too difficult, but understanding it would. It took them months to reach their tentative conclusion: if quantum mechanics described the data, then the light's polarizations didn't exist before being measured. Realism in quantum mechanics would be untenable.

On my final morning in Vienna, snow was tumbling like dryer sheets as I stared out the window of the IQOQI waiting to speak again with Zeilinger. Suddenly, there was a great flash of lightning and a long roll of thunder as snow continued to fall. I turned around to no one, and Zeilinger's assistant appeared. He now had time to talk.

Though less robust and more intimidating, Zeilinger bears a slight resemblance to the American Kris Kringle. Born in 1945, he is tall and stout, with a beard and white mane of hair. He wears tailored jackets, though he insists he is a hands-on kind of guy.

As a student in Vienna in the 1960s, Zeilinger never attended a single course in quantum mechanics, which may help to explain the way he has investigated it since — with the zeal of a late convert. In the past decade or so, Zeilinger and his many collaborators were the first to teleport light, use quantum cryptography for a bank transaction (with optical fibers in the sewers of Vienna), realize a one-way quantum computer, and achieve entanglement over large distances through the air, first across the Danube River and then between two of the Canary Islands. Zeilinger's work had also previously shown the greatest distinction between quantum mechanics and local realism.

Zeilinger's office is large and sparsely decorated. A few books lean on a lengthy, glass-fronted bookshelf. As he spoke, Zeilinger reclined in a black chair, and I leaned forward on a red couch. "Quantum mechanics is very fundamental, probably even more fundamental than we appreciate," he said. "But to give up on realism altogether is certainly wrong. Going back to Einstein, to give up realism about the moon, that's ridiculous. But on the quantum level we do have to give up realism."

With eerie precision, the results of Gröblacher's weekend experiments had followed the curve predicted by quantum mechanics. The data defied the predictions of Leggett's model by three orders of magnitude. Though they could never observe it, the polarizations truly did not exist before being measured. For so fundamental a result, Zeilinger and his group needed to test quantum mechanics again. In a room atop the IQOQI building, another Ph.D. student, Alessandro Fedrizzi, recreated the experiment using a laser found in a Blu-ray disk player.

Leggett's theory was more powerful than Bell's because it required that light's polarization be measured not just like the second hand on a clock face but over an entire sphere. In essence, there were an infinite number of clock faces on which the second hand could point. For the experimenters this meant that they had to account for an infinite number of possible measurement settings. So Zeilinger's group rederived Leggett's theory for a finite number of measurements. There were certain directions the polarization would more likely face in quantum mechanics. This test was more stringent. In mid-2007 Fedrizzi found that the new realism model was violated by eighty orders of magnitude; the group was even more assured that quantum mechanics was correct.

Leggett agrees with Zeilinger that realism is wrong in quantum mechanics, but when I asked him whether he now believes in the theory, he answered only "no" before demurring, "I'm in a small minority with that point of view and I wouldn't stake my life on it." For Leggett there are still enough loopholes to disbelieve. I asked him what could finally change his mind about quantum mechanics. Without hesitation, he said sending humans into space as detectors to test the theory. In space there is enough distance to exclude communication between the detectors (humans), and the lack of other particles should allow most entangled photons to reach the detectors unimpeded. Plus each person can decide independently which photon polarizations to measure. If Leggett's model were contradicted in space, he might believe. When I mentioned this to Zeilinger, he said, "That will happen someday. There is no doubt in my mind. It is just a question of technology." Alessandro Fedrizzi had already shown me a prototype of a realism experiment he is hoping to send up in a satellite. It's a heavy metallic slab the size of a dinner plate.

On Markus Aspelmeyer's desk there are three tall, empty boxes of Veuve Clicquot. Experimentalists at the IQOQI receive champagne for exceptional results, and on one of the boxes is written congratulations for Aspelmeyer's initiation of the realism test. Časlav Brukner, who helped with the theory, keeps a squat box of Chinese plum wine on his desk facing Aspelmeyer's. When I asked about the wine, thinking it the theorists' complementary tradition, he laughed and said he just needed a counterbalance. Brukner has an easy manner and has been with Zeilinger's group almost continuously since arriving in Austria in 1991 after leaving then-Yugoslavia.

Last year Brukner and his student Johannes Kofler decided to figure out why we do not perceive the quantum phenomena around us. If quantum mechanics holds universally for atoms, why do we not see directly its effects in bulk?

Most physicists believe that quantum effects get washed out when there are a large number of particles around. The particles are in constant interaction, and their environment serves to "decohere" the quantum world — eliminate superpositions — to create the classical one we observe. Quantum mechanics has within it its own demise, and the process is too rapid to ever see. Zeilinger's group, which has tested decoherence, does not believe there is a fundamental limit on the size of an object to observe superposition. Superpositions should exist even for objects we see, similar to the infamous example of Schrödinger's cat. In fact, Gröblacher now spends his nights testing larger-scale quantum mechanics in which a small mirror is humanely substituted for a cat.

Brukner and Kofler had a simple idea. They wanted to find out what would happen if they assumed that a reality similar to the one we experience is true — every large object has only one value for each measurable property that does not change. In other words, you know your couch is blue, and you don't expect to be able to alter it just by looking. This form of realism, "macrorealism," was first posited by Leggett in the 1980s.

Late in 2007 Brukner and Kofler showed that it does not matter how many particles are around or how large an object is; quantum mechanics always holds true. The reason we see our world as we do is because of what we use to observe it. The human body is a just barely adequate measuring device. Quantum mechanics does not always wash itself out, but to observe its effects for larger and larger objects we would need more and more accurate measurement devices. We just do not have the sensitivity to observe the quantum effects around us. In essence we do create the classical world we perceive, and as Brukner said, "There could be other classical worlds completely different from ours."

Zeilinger and his group have only just begun to consider the grand implications of all their work for reality and our world. Like others in their field, they had focused on entanglement and decoherence to construct our future information technology, such as quantum computers, and not for understanding reality. But the group's work on these kinds of applications pushed up against quantum mechanics' foundations. To repeat a famous dictum, "All information is physical." How we get information from our world depends on how it is encoded. Quantum mechanics encodes information, and how we obtain this through measurement is how we study and construct our world.

I asked Dr. Zeilinger about this as I was about to leave his office. "In the history of physics, we have learned that there are distinctions that we really should not make, such as between space and time . . . It could very well be that the distinction we make between information and reality is wrong. This is not saying that everything is just information. But it is saying that we need a new concept that encompasses or includes both." Zeilinger smiled as he finished: "I throw this out as a challenge to our philosophy friends."

A few weeks later I was looking around on the IQOQI Web site when I noticed a job posting for a one-year fellowship at the institute. They were looking for a philosopher to collaborate with the group.

OLIVER SACKS

Darwin and the Meaning of Flowers

FROM *The New York Review of Books*

WE ALL KNOW THE CANONICAL STORY of Charles Darwin: the twenty-two-year-old embarking on the *Beagle,* going to the ends of the Earth; Darwin in Patagonia; Darwin on the Argentine pampas (managing to lasso the legs of his own horse); Darwin in South America, collecting the bones of giant extinct animals; Darwin in Australia — still a religious believer — startled at his first sight of a kangaroo ("surely two distinct Creators must have been at work"). And, of course, Darwin in the Galápagos, observing how the finches were different on each island, starting to experience the seismic shift in understanding how living things evolve that, a quarter of a century later, would result in the publication of *On the Origin of Species.*

The story climaxes here, with the publication of the *Origin* in November 1859, and has a sort of elegiac postscript: a vision of the older and ailing Darwin, in the twenty-odd years remaining to him, pottering around his gardens at Down House with no particular plan or purpose, perhaps throwing off a book or two, but with his major work long completed.

Nothing could be farther from the truth. Darwin remained intensely sensitive both to criticisms and to evidence supporting his theory of natural selection, and this led him to bring out no fewer than five editions of the *Origin.* He may indeed have retreated (or returned) to his garden and his greenhouses after 1859 (there were extensive grounds around Down House and five greenhouses), but for him these became engines of war, from which he would lob great missiles of evidence at the skeptics outside — descriptions of extraordinary structures and behaviors in plants very difficult to ascribe to special Creation or Design — a mass of evidence for evolution and natural selection even more overwhelming than that presented in the *Origin*.

Strangely, even Darwin scholars pay relatively little attention to this botanical work, even though it encompassed six books and seventy-odd papers. Thus Duane Isely, in his 1994 book *One Hundred and One Botanists*, writes that while

> more has been written about Darwin than any other biologist who ever lived . . . he is rarely presented as a botanist . . . The fact that he wrote several books about his research on plants is mentioned in much Darwinia but it is casual, somewhat in the light of "Well, the great man needs to play now and then."

Even now, as we approach the two hundredth anniversary of Darwin's birth and the hundred and fiftieth of the *Origin*, this is still very much the case, and it was with this in mind that the New York Botanical Garden recently launched an exhibition called "Darwin's Garden: An Evolutionary Adventure." This contained not only reconstructions of the gardens at Down and many of the actual experiments Darwin set up there, but a mass of rare books, papers, letters, and drawings.

Darwin had always had a special, tender feeling for plants and a special admiration, too ("it has always pleased me to exalt plants in the scale of organised beings," he wrote in his autobiography). He grew up in a botanical family — his grandfather, Erasmus Darwin, had written a long, two-volume poem called *The Botanic Garden*, and Charles himself grew up in a house whose extensive gardens were filled not only with flowers but with a variety of apple trees crossbred for increased vigor. As a university student at Cambridge, the only lectures Darwin consistently attended were those of the botanist J. S. Henslow, and it was Henslow, recognizing the extraordinary qualities of his student, who recommended him for a position on the *Beagle*.

It was to Henslow that Darwin wrote very detailed letters full of observations about the fauna and flora and geology of the places he visited. (These letters, when printed and circulated, were to make Darwin famous in scientific circles even before the *Beagle* returned to England.) And it was for Henslow that Darwin, in the Galápagos, made a careful collection of all the plants in flower and noted how different islands in the archipelago could often have different species of the same genus. This was to become a crucial piece of evidence for him as he thought about the role of geographical divergence in the origin of new species.

Indeed, as David Kohn points out in his splendid essay, "Darwin's Galápagos plant specimens, numbering well over 200, constitute the single most influential natural history collection of live organisms in the entire history of science . . . They also would turn out to be Darwin's best documented example of the evolution of species on the islands." (The birds Darwin collected, by contrast, were not always correctly identified or labeled with their island of origin; and it was only on his return to England that these — supplemented by the specimens collected by his shipmates — were sorted out by the ornithologist John Gould.)

Darwin had become close friends with two botanists — Joseph Dalton Hooker, at Kew Gardens, and Asa Gray at Harvard. Hooker had become his confidant in the 1840s — the only man to whom he showed the first draft of his work on evolution; and Asa Gray was to join the inner circle in the 1850s. Thus he would write to them both with increasing enthusiasm about "*our* theory."

Though Darwin was happy to call himself a geologist (he wrote three geological books based on his observations during the voyage of the *Beagle* and conceived a strikingly original theory on the origin of coral atolls, which was confirmed experimentally only in the second half of the twentieth century), he always insisted that he was not a botanist. One reason was that botany had (despite a precocious start in the early eighteenth century with Stephen Hales's *Vegetable Statics,* a book full of fascinating experiments on plant physiology) remained almost entirely a descriptive and taxonomic discipline — plants were identified, classified, and named, but not investigated. Darwin, by contrast, himself was preeminently an investigator, concerned with the "how" and "why" of plant structure and behavior, not just the "what."

Botany was not a mere avocation or hobby for Darwin, as it was for so many in the Victorian age; the study of plants was always infused, for him, with theoretical purpose, and the theoretical purpose had to do with evolution and natural selection. It was, as his son Francis wrote, "as though he were charged with theorising power ready to flow into any channel on the slightest disturbance, so that no fact, however small, could avoid releasing a stream of theory." And the flow went both ways; Darwin himself often said that "no one could be a good observer unless he was an active theoriser."

In the eighteenth century, the Swedish scientist Linnaeus had shown that flowers had sexual organs (pistils and stamens) and indeed had based his classifications on these. But it was almost universally believed that flowers were self-fertilized — why else would each flower contain both male and female organs? Linnaeus himself made merry with the idea, portraying a flower with nine stamens and one pistil as a bedchamber in which a maiden was surrounded by nine lovers. A similar conceit appeared in the second volume of Darwin's grandfather's book, *The Botanic Garden,* titled "The Loves of Plants." This was the atmosphere in which the younger Darwin grew up.

But within a year or two of his return from the *Beagle,* Darwin felt forced, on theoretical grounds, to question the idea of self-fertilization. In an 1837 notebook, he wrote, "Do not plants which have male and female organs together yet receive influence from other plants?" If plants were ever to evolve, he reasoned, cross-fertilization was crucial — otherwise, no changes, no modifications could ever occur, and the world would be stuck with a single, self-reproducing plant instead of the extraordinary range of species it actually had. In the early 1840s, Darwin started to test his theory, dissecting a variety of flowers (azaleas and rhododendrons among them) and demonstrating that many of these had structural devices for preventing or minimizing self-pollination.

But it was only after *On the Origin of Species* was published in 1859 that Darwin could turn his full attention to plants. And where his early work was primarily as an observer and a collector, experiments now became his chief way of obtaining new knowledge.

He had observed, as others had, that primrose flowers came in two different forms: a "pin" form with a long style — the female part of the flower — and a "thrum" form with a short style. These differences were thought to have no particular significance. But Darwin suspected otherwise, and, examining bunches of primroses that his children brought him, he found that the ratio of pins to thrums was exactly one to one.

Darwin's imagination was instantly aroused: a one-to-one ratio was what one might expect of species with separate males and females — could it be that the long-styled flowers, though hermaphrodites, were in the process of becoming female flowers and the short-styled ones male flowers? Was he actually seeing intermediate forms, evolution in action? It was a lovely idea, but it did not hold up, for the short-styled flowers, the putative males, produced as much seed as the long-styled, "female" ones. Here (as his friend T. H. Huxley would have put it) was "the slaying of a beautiful hypothesis by an ugly fact."

What, then, was the meaning of these different styles and their one-to-one ratio? Giving up theorizing, Darwin turned to experiment. Painstakingly, he tried acting as a pollinator himself, lying face down on the lawn and transferring pollen from flower to flower: long-styled to long-styled, short-styled to short-styled, long-styled to short-styled, and vice versa. When seeds were produced, he collected and weighed them, and found that the richest crop of seeds came from the crossbred flowers. He concluded that heterostyly, in which plants have styles of different length, was a special device that had evolved to facilitate outbreeding — and that crossing increased the number and vitality of seeds (he called this "hybrid vigour"). Darwin later wrote, "I do not think anything in my scientific life has given me so much satisfaction as making out the meaning of the structure of these plants."

Although this subject remained a special interest of Darwin's (he published a book on it in 1877, *The Different Forms of Flowers on Plants of the Same Species*), his central concern was how flowering plants adapted themselves to using insects as agents for their own fertilization. It was well known that insects were attracted to certain flowers, visited them, and could emerge from blossoms covered with pollen. But no one had thought this was of much importance, since it was assumed that flowers were self-pollinated.

Darwin had already become very suspicious of this by 1840, and in the 1850s he set five of his children to work, plotting the flight routes of male humble bees. He especially admired the native orchids that grew in the meadows around Down, so he started with them. Then, with the help of friends and correspondents who sent him orchids to study, and especially Hooker, who was now director of Kew Gardens, he extended his studies to tropical orchids of all kinds.

The orchid work moved quickly and well, and in 1862 Darwin was able to send his manuscript to the printers. The book had a typically long and explicit Victorian title, *The Various Contrivances by Which Orchids Are Fertilised by Insects*. His intentions, or hopes, were made clear in its opening pages:

> In my volume "On the Origin of Species" I gave only general reasons for the belief that it is an almost universal law of nature that the higher organic beings require an occasional cross with another individual . . . I wish here to show that I have not spoken without having gone into details . . . This treatise affords me also an opportunity of attempting to show that the study of organic beings may be as interesting to an observer who is fully convinced that the structure of each is due to secondary laws, as to one who views every trifling detail of structure as the result of the direct interposition of the Creator.

Here, in no uncertain terms, Darwin is throwing down the gauntlet, saying, "Explain *that* better — if you can."

Darwin interrogated orchids, interrogated flowers, as no one had ever done before, and in his orchid book he provided enormous detail, far more than is to be found in the *Origin*. This was not because he was pedantic or obsessional, but because he felt that every detail was potentially significant. It is sometimes said that God is in the details, but for Darwin it was not God but natural selection, acting over millions of years, which shone out from the details, details which were unintelligible, senseless, except in the light of history and evolution. His botanical researches, his son Francis wrote,

> supplied an argument against those critics who have so freely dogmatised as to the uselessness of particular structures, and as to the consequent impossibility of their having been developed by means of natural selection. His observations on Orchids enabled him to say: "I can show the meaning of some of the apparently meaningless ridges and horns; who will now venture to say that this or that structure is useless?"

In a 1793 book entitled *The Secret of Nature Revealed in the Structure and Fertilization of Flowers*, the German botanist Christian Konrad

Sprengel, a most careful observer, had noted that bees laden with pollen would carry it from one flower to another. Darwin always called this a "wonderful" book. But Sprengel, though he drew close, missed the final secret, because he was still wedded to the Linnaean idea of flowers as self-fertilizing — and thought of flowers of the same species as essentially identical. It was here that Darwin made a radical break and cracked the secret of flowers, by showing that their special features — the various patterns, colors, shapes, nectars, and scents by which they lured insects to flit from one plant to another, and the devices which ensured that the insects would pick up pollen before they left the flower — were all "contrivances," as he put it; they had all evolved in the service of cross-fertilization.

What had once been a pretty picture of insects buzzing about brightly colored flowers now became an essential drama in life, full of biological depth and meaning. The colors and smells of flowers were adapted to insects' senses. While bees are attracted to blue and yellow flowers, they ignore red ones, because they are red-blind. On the other hand, their ability to see beyond the violet is exploited by flowers that use ultraviolet markings — the so-called honeyguides, which direct bees to their nectaries. Butterflies, with good red vision, fertilize red flowers but may ignore the blue and violet ones. Flowers pollinated by night-flying moths tend to lack color but to exude their scents at night. And flowers pollinated by flies, which live on decaying matter, may mimic the (to us) foul smells of putrid flesh.

It was not just the evolution of plants but the coevolution of plants and insects that Darwin illuminated for the first time. Thus natural selection would ensure that the mouthparts of insects matched the structure of their preferred flowers — and Darwin took special delight in making predictions here. Examining one Madagascan orchid with a nectary nearly a foot long, he predicted that a moth would be found with a proboscis long enough to probe its depths; such a moth was finally discovered, decades after his death.

The *Origin* was a frontal assault (delicately presented though it was) on creationism, and while Darwin had been careful to say little in the book about human evolution, the implications of his theory were perfectly clear. It was especially the idea that man could be regarded as a mere animal — an ape — descended from other animals that had provoked outrage and ridicule. But for most people, plants were a different matter — they neither moved nor felt; they inhabited a kingdom of their own, separated from the animal kingdom by a great gulf. The evolution of plants, Darwin sensed, might seem less relevant, or less threatening, than the evolution of animals, and so more accessible to calm and rational consideration. Indeed, he wrote to his friend Asa Gray, "No one else has perceived that my chief interest in my orchid book, has been that it was a 'flank movement' on the enemy." Darwin was never belligerent, like his "bulldog" Huxley, but he knew that there was a battle to wage, and he was not averse to military metaphors.

It is, however, not militancy or polemic that shines out of the orchid book; it is sheer joy, delight in what he was seeing. This delight and exuberance burst out of his letters:

> You cannot conceive how the Orchids have delighted me . . . What wonderful structures! . . . The beauty of the adaptation of parts seems to me unparalleled . . . I was almost mad at the wealth of Orchids. . . . One splendid flower of Catasetum, the most wonderful Orchid I have seen . . . Happy man, he [who] has actually seen crowds of bees flying round Catasetum, with the pollinia sticking to their backs! . . . I never was more interested in any subject in all my life than in this of Orchids.

The fertilization of flowers engaged Darwin to the end of his life, and the orchid book was to be followed, nearly fifteen years later, by a more general book, *The Effects of Cross and Self Fertilisation in the Vegetable Kingdom*.

But plants have to survive, flourish, and find (or create) niches in the world, if they are ever to reach the point of reproduction. Thus Darwin's equal interest in the devices and adaptations by which plants survived and their varied and sometimes astonishing lifestyles, which included sense organs and motor powers akin to those of animals.

In 1860, during a summer holiday, Darwin first encountered and became enamored of insect-eating plants — and this started a series of investigations that would culminate, fifteen years later, in the publication of *Insectivorous Plants*. This volume has an easy, companionable style, and starts, like most of his books, with a personal recollection:

> I was surprised by finding how large a number of insects were caught by
> the leaves of the common sun-dew *(Drosera rotundifolia)* on a heath in
> Sussex . . . On one plant all six leaves had caught their prey . . . Many
> plants cause the death of insects . . . without thereby receiving, as far as
> we can perceive, any advantage; but it was soon evident that Drosera was
> excellently adapted for the special purpose of catching insects.

The idea of adaptation was always in Darwin's mind, and one look
at the sundew showed him that these were adaptations of an entirely novel kind, for *Drosera*'s leaves not only had a sticky surface
but were covered with delicate filaments (Darwin called them "tentacles") with glands at their tips. What were these for, he wondered? "If a small organic or inorganic object be placed on the
glands in the centre of a leaf," he observed, "they transmit a motor
impulse to the marginal tentacles . . . The nearer ones are first affected and slowly bend towards the centre, and then those farther
off, until at last all become closely inflected over the object." But if
the object was not nourishing, it was speedily released.

Darwin went on to demonstrate this by putting blobs of egg
white on some leaves and similar blobs of inorganic matter on others. The inorganic matter was quickly released, but the egg white
was retained and stimulated the formation of a ferment and an
acid that soon digested and absorbed it. It was similar with insects,
especially live ones. Here, without a mouth, or a gut, or nerves,
Drosera efficiently captured its prey and, using special digestive enzymes, absorbed it.

Darwin addressed not only how *Drosera* functioned but why it
had adopted so extraordinary a lifestyle: he observed that the plant
grew in bogs, in acidic soil that was relatively barren of organic material and assimilable nitrogen. Few plants could survive in such
conditions, but *Drosera* had found a way to claim this niche by absorbing its nitrogen directly from insects rather than from the
soil. Amazed by the animal-like coordination of *Drosera*'s tentacles,
which closed on its prey like those of a sea anemone, and by the
plant's animal-like ability to digest, Darwin wrote to Asa Gray: "You
are unjust on the merits of my beloved *Drosera*; it is a wonderful
plant, or rather a most sagacious animal. I will stick up for *Drosera*
to the day of my death."

And he became still more enthusiastic about *Drosera* when he
found that making a small nick in half of a leaf would paralyze just
that half, as if a nerve had been cut. The appearance of such a leaf,

he wrote, resembled "a man with his backbone broken and lower extremities paralysed." Darwin later received specimens of the Venus flytrap — a member of the sundew family — which, the moment its triggerlike hairs were touched, would clap its leaves together on an insect and imprison it. The flytrap's reactions were so fast that Darwin wondered whether electricity could be involved, something analogous to a nerve impulse. He discussed this with his physiologist colleague Burdon Sanderson and was delighted when Sanderson was able to show that electric current was indeed generated by the leaves and could also stimulate them to close. "When the leaves are irritated," Darwin recounted in *Insectivorous Plants*, "the current is disturbed in the same manner as takes place during the contraction of the muscle of an animal."

Plants are often regarded as insensate and immobile — but the insect-eating plants provided a spectacular rebuttal of this notion, and now, eager to examine other aspects of plant motion, Darwin turned to an exploration of climbing plants. (This would culminate in the publication of *On the Movements and Habits of Climbing Plants*.) Climbing was an efficient adaptation, allowing plants to disburden themselves of rigid supporting tissue and to use other plants to support and elevate them. And there was not just one way of climbing, but many. There were twining plants, leaf-climbers, and plants that climbed with the use of tendrils. These especially fascinated Darwin — it was almost, he felt, as if they had "eyes" and could "survey" their surroundings for suitable supports. "I believe, Sir, the tendrils can see," he wrote to Asa Gray. How did such complex adaptations arise?

Darwin saw twining plants as ancestral to other climbing plants, and he thought that tendril-bearing plants had evolved from these, and leaf-climbers, in turn, from tendril-bearers, each development opening up more and more possible niches — roles for the organism in its environment. Thus climbing plants had evolved over time — they had not all been created in an instant, by divine fiat. But how did twining itself start? Darwin had observed twisting movements in the stems, leaves, and roots of every plant he had examined, and such twisting movements (which he called circumnutation) could also be observed in all "lower" plants: cycads, ferns, seaweeds, too. When plants grow toward the light, they do not just thrust upward, they twist, they corkscrew, toward the light. Circumnutation, Darwin came to think, was a universal disposition of plants, and the antecedent of all other twisting movements in plants.

These thoughts, along with dozens of beautiful experiments, were set out in his last botanical book, *The Power of Movement in Plants,* published in 1880. Among the many charming and ingenious experiments he recounted was one in which he planted oat seedlings, shone light on them from different directions, and found that they always bent or twisted toward the light, even when it was too dim to be seen by human eyes. Was there (as he imagined of the tips of tendrils) a photosensitive region, a sort of "eye" at the tips of the seedling leaves? He devised little caps, darkened with India ink, to cover these, and found that they no longer responded to light. It was clear, he concluded, that when light fell on the leaf tip, it stimulated the tip to release some sort of messenger which, reaching the "motor" parts of the seedling, caused it to twist toward the light. Similarly, the primary roots (or radicles) of seedlings, which have to negotiate all sorts of obstacles, Darwin found to be extremely sensitive to contact, gravity, pressure, moisture, chemical gradients, etc. He wrote: "There is no structure in plants more wonderful, as far as its functions are concerned, than the tip of the radicle . . . It is hardly an exaggeration to say that the tip of the radicle . . . acts like the brain of one of the lower animals . . . receiving impressions from the sense-organs, and directing the several movements."

But as Janet Browne remarks, *The Power of Movement in Plants* was "an unexpectedly controversial book." Darwin's idea of circumnutation was roundly criticized. He had always acknowledged it as a speculative leap, but a more cutting criticism came from the German botanist Julius Sachs, who, in Browne's words, "sneered at Darwin's suggestion that the tip of the root might be compared to the brain of a simple organism and declared that Darwin's home-based experimental techniques were laughably defective."

However homely Darwin's techniques, though, his observations were precise and correct. His ideas of a chemical messenger being transmitted downward from the sensitive tip of the seedling to its "motor" tissue were to lead the way, fifty years later, to the discovery of plant hormones like auxins, which, in plants, play the roles that nervous systems do in animals.

Darwin had been an invalid for forty years, with an enigmatic illness that had assailed him since his return from the Galápagos. He would sometimes spend entire days vomiting, or confined to his sofa, and as he grew older, he developed heart problems, too. But his intellectual energy and creativity never wavered. He wrote a total of ten books after the *Origin,* many of which themselves went through major revisions — to say nothing of dozens of articles and innumerable letters. He continued to pursue his varied interests throughout his life. In 1877, for instance, he published a second, greatly enlarged and revised, edition of his orchid book (originally published fifteen years earlier). My friend Eric Korn, an antiquarian and Darwin specialist, tells me that he recently had a copy of this in which there was slipped the counterfoil of an 1882 postal order for two shillings and ninepence, signed by Darwin himself, in payment for a new orchid specimen. Darwin was to die in April of that year, but he was still in love with orchids and collecting them for study within weeks of his death.

His final book, *The Formation of Vegetable Mould Through the Action of Worms, with Observations on Their Habits,* published the year before his death, returned to a favorite subject — earthworms — which he had first written about more than forty years earlier. Starting from homely experiments in his own garden and extrapolating from these, Darwin brought out that these seemingly insignificant creatures — hitherto thought of, if at all, as pests — had been instrumental in altering the geography and geology of the Earth, digesting organic matter and turning it into soil. He calculated that in England alone, worms could transform more than 100 billion tons of earth within a thousand years. Darwin's book on worms was, as Robbin Moran points out, "about how slow, gradual processes working over a long time could have a great cumulative effect. Given enough time, worms can bury a house. The parallel to natural selection is obvious."

Natural beauty, for Darwin, was not just aesthetic; it always reflected function and adaptation at work. Orchids were not just ornamental, to be displayed in a garden or a bouquet; they were wonderful contrivances, examples of nature's imagination, natural selection, at work. Flowers required no Creator, but were wholly intelligible as products of accident and selection, of tiny incremental changes extending over hundreds of millions of years. This, for Darwin, was the meaning of flowers, the meaning of all adaptations, plant and animal, the meaning of natural selection.

It is often felt that Darwin, more than anyone, banished "meaning" from the world — in the sense of any overall divine meaning or purpose. There is indeed no design, no plan, no blueprint in Darwin's world; natural selection has no direction or aim, nor any goal to which it strives. Darwinism, it is often said, spelled the end of teleological thinking. And yet, his son Francis writes:

> One of the greatest services rendered by my father to the study of Natural History is the revival of Teleology. The evolutionist studies the purpose or meaning of organs with the zeal of the older Teleologist, but with far wider and more coherent purpose. He has the invigorating knowledge that he is gaining not isolated conceptions of the economy of the present, but a coherent view of both past and present. And even where he fails to discover the use of any part, he may, by a knowledge of its structure, unravel the history of the past vicissitudes in the life of the species. In this way a vigour and unity is given to the study of the forms of organised beings, which before it lacked.

And this, Francis suggests, was "effected almost as much by Darwin's special botanical work as by the *Origin of Species*."

By asking why, by seeking meaning (not in any final sense, but in the immediate sense of use or purpose), Darwin found in his botanical work the strongest evidence for evolution and natural selection. And in doing so, he transformed botany itself from a purely descriptive discipline into an evolutionary science. Botany, indeed, was the first evolutionary science, and Darwin's botanical work was to lead the way to all the other evolutionary sciences — and to the insight, as Theodosius Dobzhansky put it, that "nothing in biology makes sense except in the light of evolution."

Darwin spoke of the *Origin* as "one long argument." His botanical books, by contrast, were more personal and lyrical, less systematic in form, and they secured their effects by demonstration, not argument. According to Francis Darwin, Asa Gray observed that if the orchid book "had appeared before the *Origin,* the author would have been canonized rather than anathematized by the natural theologians."

Linus Pauling records in an autobiographical essay that he read the *Origin* when he was ten. I was not that precocious and could not have followed its "one long argument" at that age. But I had an intimation of Darwin's vision of the world in our own garden — a garden which, on summer days, was full of flowers and bees buzzing from one flower to another. It was my mother, botanically inclined, who explained to me what the bees were doing, their legs yellow with pollen, and how they and the flowers depended on each other.

While most of the flowers in the garden had rich scents and colors, we also had two magnolia trees, with huge but pale and scentless flowers. The magnolia flowers, when ripe, would be crawling with tiny insects, little beetles. Magnolias, my mother explained, were among the most ancient of flowering plants and had appeared nearly a hundred million years ago, at a time when "modern" insects like bees had not yet evolved, so they had to rely on a more ancient insect, a beetle, for pollination. Bees and butterflies, flowers with colors and scents, were not preordained, waiting in the wings — and they might never have appeared. They would develop together, in infinitesimal stages, over millions of years. The idea of a world without bees or butterflies, without scent or color, affected me with a sense of awe.

The notion of such vast eons of time, and the power of tiny, undirected changes which by their accumulation could generate new worlds — worlds of enormous richness and variety — was intoxicating. Evolutionary theory provided, for many of us, a sense of deep meaning and satisfaction that belief in a Divine Plan had never achieved. The world that presented itself to us became a transparent surface, through which one could see the whole history of life. The idea that it could have worked out differently, that dinosaurs might still be roaming the Earth or that human beings might never have evolved, was a dizzying one. It made life seem all the more precious, and a wonderful, ongoing adventure ("a glorious accident," as Stephen Jay Gould called it) — not fixed or predetermined, but always susceptible to change and new experience.

Life on our planet is several billion years old, and we literally embody this deep history in our structures, our behaviors, our instincts, our genes. We humans retain, for example, the remnants of gill arches, much modified, from our fishy ancestors — and even the neural systems that once controlled gill movement. As Darwin wrote in *The Descent of Man,* "Man still bears in his bodily frame the indelible stamp of his lowly origin."

In 1837, in the first of many notebooks he was to keep on "the species problem," Darwin sketched a tree of life. Its brachiating shape, so archetypal and potent, reflected the balance of evolution and extinction. Darwin always stressed the continuity of life, how all living things are descended from a common ancestor, and how we are in this sense all related to each other. So humans are related not only to apes and other animals, but to plants too. (Plants and animals, we know now, share 70 percent of their DNA.) And yet, because of that great engine of natural selection — variation — every species is unique and each individual is unique, too.

The tree of life shows at a glance the antiquity and the kinship of all living organisms, and how there is (as Darwin originally called evolution) "descent with modification" at every juncture. It shows too that evolution never stops, never repeats itself, never goes backward. It shows the irrevocability of extinction — if a branch is cut off, a particular evolutionary path is lost forever.

I rejoice in the knowledge of my biological uniqueness and my biological antiquity and my biological kinship with all other forms of life. This knowledge roots me, allows me to feel at home in the natural world, to feel that I have my own sense of biological meaning, whatever my role in the cultural, human world. And although animal life is far more complex than vegetable life, and human life far more complex than the life of other animals, I trace back this sense of biological meaning to Darwin's epiphany on the meaning of flowers, and to my own intimations of this in a London garden, nearly a lifetime ago.

MARK A. SMITH

Animalcules and Other Little Subjects

FROM *Isotope*

EVERY SO OFTEN, when I get a chance to step away from the routines of the teaching life, I like to spend some time below the world's surface and remind myself just how uncanny and improbable this thing called life really is. It's a good exercise for creatures of habit such as ourselves, who too often neglect worlds other than our own, and forget, too, that there are worlds within worlds nearly everywhere we turn. So I remind myself to go lower, go deeper.

A pickle jar filled with pond water will do, along with an old microscope and a spare afternoon. That's what I have in mind today as I walk down to the fecund ten-acre pond that sits near my house. I dip the jar at the water's edge, and everything I need pours right in. To complicate matters I add a drowned cottonwood leaf grown fuzzy with detritus, a waving clump of grass-green algae plucked from a sunken log, and some mud and grit from the bottom. I overturn pebbles to find the flatworms known as planarians clinging below — the gray ones today, not the orange ones I like so well. In they go, each piece a different environment, a different ecosystem. This bright green algae might be the Amazon: tangled, fast-paced, diverse. A dead leaf is like a rugged steppe with herds of fleet grazers and a few roving predators. A rounded pebble reminds me of the Appalachian mountains where sluggish brown bears reside. At the microscopic level the idea of wilderness gets a new lease on life. With the microcosm assembled, I head back to the house and plunk the jar down beside my scope out here on the back porch. I get out the slides and cover slips, the forceps, the eyedropper. Minimal equipment really, nothing fancy. Then I sit back with a glass of iced tea and wait for the silt and the worlds to settle. The animalcules slowly reclaim their niches, working toward the surface, the water column, the sediment — wherever they prefer to dwell.

Antony van Leeuwenhoek first used the term "animalcules" in 1674 to describe what he saw with his microscopes. He was one of my heroes as a child, along with Leonardo da Vinci and Johnny Bench, an odd mix of heroes. Leeuwenhoek was himself an odd mix, a Dutch draper, surveyor, and wine-gauger by trade, but his lifelong passion was making and using microscopes. He made hundreds of them, each with a single tiny lens mounted between two metal plates. One of his lenses was ground and polished from a grain of sand. Another magnified up to 266 times. The subject was mounted on a pin and worked by screws until it focused before the lens; then the simple contraption, just an inch by two inches overall, was held up to the light. Leeuwenhoek was the very first to see bacteria, the first to describe red blood cells, the first to find rotifers, hydra, *Volvox* — all these strange glories utterly new to the human imagination and all because of his ingenious little scopes. Of all the microscopes he made, however, only nine survive. It seems he liked to craft them of silver, sometimes even gold, so the rest were melted down for a cheaper kind of wealth than they yielded intact.

The loss is sad because good microscopes are nearly immortal. My Bausch & Lomb, for instance, had three owners before me and must be pushing sixty years of age. It is beautifully machined from cast iron and brass, enameled all in glossy black, bristling with rhodium-plated knobs — a pair each for coarse and fine focusing, two more to work the mechanical stage, yet another to raise and focus the substage condenser. The aura of the thing is solid and purposeful. Precision exudes from every rack and pinion. A trio of keenly polished objective lenses are set like jewels in the revolving turret.

While working these controls I am often as caught up and busy as a child. At the helm of this machine I make course corrections across the microcosm of the slide. Time fades, and afternoons pass in rapt involvement the way they did when I was young. I suppose there's a boyish quality to all this, but I never get enough of it. After a few hours I come away with my left eye, my viewing eye, just about blinded. I grope through a life-size, lumbering world where, compared to the bright delineation of the scope, everything seems temporarily crude and rounded off.

Now, as the sediment slowly clears, and even without the microscope's aid, I can see copepods, ostracods, and cladocerans all zinging around in the pickle jar. These are all crustaceans, and most are just a millimeter or two in size. Today *Cyclops* outnumbers the rest, a copepod with one ruby-red eye set in an intricate crystalline body shaped something like a shrimp. They flit and jerk through the water doing a spasmodic breaststroke with their antennules. The females carry paired ovisacs stuffed with developing young. I also find about a dozen water fleas, the cladocerans known as *Daphnia*. A large female gravid with young hangs head down on the side of the jar. Quick and alert, she nearly escapes the eyedropper, but I get her into a welled slide, where she kicks around in circles before settling down to feed. Her appendages run in place, rhythmically fanning particles of food to her mouth.

This *Daphnia* has everything required for life as we humans know it — eyes, heart, brain, muscles, stomach, arms, and legs — all packed into a shell two millimeters long. Everything but the eye is translucent: the carapace, the antennae, the appendages, the blood itself. I watch her heart beating and the blood flowing throughout, as clear as distilled water but flecked with a silver confetti of blood cells, which spout forward past the eye and branch to either side of the rounded carapace like a gentle fountain. The cells glide over the large clear muscles that work the antennae for swimming (a strong swimmer, she has the biceps of a bodybuilder). In her dorsal brood pouch, just behind the heart, I see all her young. She has a dozen or more in there, and at this stage of development they are bean-shaped things with two dark eyes like tiny goblins. They squirm and jostle like children strapped in car seats and then, like their mother, they settle down once more.

This mother's compound eye and nervous system must register the microscope looming starkly above her — but how? Twenty lenses compose the eye, yielding twenty views of a shadowy alien and his harshly lit instruments. An alien abduction, perhaps. Sucked up the dropper like a tractor beam and coldly examined on this bright stage. I wonder if fear is a part of her awareness. Does her heart race the way mine would? And what would be the mean-ing of that minute, half-formed fear? Of what significance is her maternal struggle in this drop of water — one drop amid ten thou-sand drops in the jar, one jar dipped randomly from a pond of a trillion jars, one pond so small the mapmakers mostly leave it off the maps? I have no answer. She is a diamond chip brought to life, and I am a little stunned by it all. When I tip the slide back into the jar she is there again, suspended and crystalline. Only the tiny, dark eye gives her presence away.

A little deeper now. In a drop from the surface, with unaided eye, I see only a speck of detritus and a white mote or two gliding slowly beneath the cover slip. A quiet-looking drop, but under the microscope at 50x it becomes a blizzard of life, with hundreds of tiny ciliates and flagellates swirling and spinning like snowflakes. In a fresh sample there is often this initial shock of activity. The drop looks quiet, sometimes dead to the naked eye, but under the scope it hums with one-celled life. Nothing sits still, of course: a drop is a dozen city blocks for the inhabitants. The animalcules are swim-ming, eating, dividing, and I chase them around like a paparazzo.

At 100x I watch a common ciliate known as *Euplotes* feeding in and around the detritus. These are flattened, bug-shaped cells that walk remarkably well on stiff, hairlike cirri. They are festooned with fringes and collars of specialized cilia beating in synchrony like chase lights on a Las Vegas marquee. Here they crawl busily over the detritus like sow bugs, first one way, then another, doing an about-face — whatever it takes to find food. Just one cell, this *Euplotes,* and barely one-tenth of a millimeter in length, but it scur-ries with purpose and aplomb in this world.

Farther on looms the much larger *Stentor,* attached to the de-tritus like the long, tapered bell of a trumpet. A spiral of cilia surrounds the bell, creating a vortex that draws food inward to the mouth. Fully extended, *Stentor* reaches a millimeter or two in length and can be visible to the naked eye. In a culture they crowd the jar's surface like razor stubble. They are feeding voraciously here in this drop, stretched out and writhing like dark blue funnel clouds, but when one of them swims off to greener pastures, it con-tracts into a plump shape and motors away like a cumbersome blimp.

In another drop vacuumed from the surface of the leaf, I find a little amoeba. Slow and grainy, it pours and oozes across the slide like transparent lava. Genus *Mayorella,* I suppose, but someday I hope to find *Chaos,* one of the largest amoebas and something of a rarity. None other than Carolus Linnaeus, the founder of taxonomy, supplied the name. He had a lifelong passion for ordering the natural world, but this disorderly organism must have given him fits, because in the end he simply called it *Chaos chaos. Difflugia* is here as well, a curious amoeba that makes a little shell or "test" for itself out of microscopic grit. The shell looks just like crushed fragments of brown glass glued precisely into the shape of a vase, and as I watch, a shy little pseudopod creeps out from under the vase's lip. *Arcella* is another common testate amoeba that forms a transparent, caramel-colored shell shaped rather like a bowl. They lie scattered about the slide like bright pennies on a sidewalk. When *Arcella* divides, it first makes a second shell, into which the daughter cell is born. These are single cells making snug little homes for themselves and providing the same for their offspring. I smile at how utterly ingrained and universal these domestic activities turn out to be.

Time for a little break. I look up from the scope and stand to stretch. Already my viewing eye is dulled. Already the yard beyond the back porch looks huge and messy and cobbled; the flowers seem ragged compared to the bright world beneath the lenses. I slip inside the house for more iced tea, then take some notes about all I've seen.

I have trouble explaining my satisfaction with the animalcules. When I think of the endless activity that takes place in the microcosm, all utterly beyond my ordinary comprehension, it makes me smile. For some reason I often feel calm and reassured afterward, perhaps because I realize how much room remains for *more.* Maybe astronomers experience something similar when viewing a distant nebula. A drastic expansion of context is involved, some bold correction to the borders of being, some small (and often temporary) progress toward weighing the self in the scales of the universe. And yet for me, at any rate, the telescope will never be as intriguing as the microscope. It's the intimacy, I think, the immediacy of the animalcules that gets me. Most of all it's the life. With so much living dazzle right here under my boot soles, under my fingernails, within my pith and spit, I say why bother with something as remote as a dying star on the far side of the galaxy?

Into the jungle now. With tweezers I pull off a bit of grass-green algae and discover a menagerie of little chalice-shaped organisms perched like birds and unfolding like flowers. *Vorticella* is here, a well-known ciliate resembling a crystal wine glass on a long, thready stem. Some go it alone, while others huddle in bright bouquets. Tethered by their stalks, they venture far from their moorings and feed, as all ciliates do, with a whirl of cilia, but every so often they contract violently, the whole length of their tether coiling instantly like a telephone cord. Over and over, they repeat this simple motion; like all protozoans, they are tireless. Around the turn of the last century, a curious researcher watched *Vorticella* for five days straight and never once saw it rest. These algae jungles are also the haunts of suctorians, heliozoans, *Zoothamnium*, *Spirogyra*, and *Opercularia*. The colorful names barely hint at the profusion and utterly fail to convey the gaudy carnival of life at this scale.

In most of these drops I've seen the ubiquitous rotifers, the "wheel animalcules," as they have long been called. Although they are metazoans and have about a thousand fused cells, rotifers are rarely bigger than the one-celled protozoans they live beside. With more than 1,800 species they are so unlike anything else on the planet that they get a phylum all to themselves. (By way of comparison, we humans share the phylum Chordata with everything from sea squirts to elephants.) Stunningly detailed, often agile and quick, most rotifers have two ciliated wheels used for both swimming and eating. Within the mouth a pair of hook-shaped jaws called the mastax grinds up the food and sends it on to the stomach. I find five or six rotifer species in the pond today — an outrageous assortment. There's a lumpish little slug of a rotifer with telescoping tail sections inchworming across the slide. It stops, grips the slide with its foot, unfurls its wheels and sets them in motion to feed. And here's a little block-headed loricate rotifer with pairs of spines protruding front and rear. It folds up its "toe" like an airplane's landing gear and, with ciliated wheels for propellers, hovers away. And here's my favorite, a rotifer I look for every time. It's a sleek, swift rotifer from the genus *Trichocerca* sporting one bright red eye, a sharp rattail, and twin fins running down its back. I find it raking and strafing through swarms of tiny flagellates, back and forth like a vacuum cleaner yet as agile as a dolphin.

Some of these rotifers are available commercially (especially *Brachionus calyciflorus* and *B. plicatilis*) under trade names such as Rotox and Rotoxkit. They are used to test waters for contamination by pesticides or industrial wastes. You hatch the encysted rotifers, put them into little wells along with your water samples, and after twenty-four hours you count the living. As sentinels of toxicity, when the rotifers die it's time to worry. Rotifers are hardy beasts and survive well in these tests, often outliving other test subjects such as minnows and *Daphnia*. In the past I have found rotifers living in sewage-laden water that would have killed a minnow in minutes, but then again I remember test results from a Tennessee Valley Authority reservoir that killed 99.9 percent of them within the day.

In the late nineteenth century, long before anyone tried to trademark a rotifer, C. T. Hudson and P. H. Gosse spent thirty years studying and cataloging their bewildering variety. In the preface to their wonderful two-volume treatise called *The Rotifera; or Wheel Animalcules,* Hudson writes:

> It is so natural to recommend one's own favourite pursuit that the recommendation often carries but little weight; and yet there is much to be said in favour of the study of Rotifera, that cannot be gainsaid. They are to be found almost everywhere; they cost nothing; they require neither expensive lenses nor an elaborate apparatus; they tempt us to explore the country, and to take pleasant walks; they are beautiful themselves; and they suggest all kinds of difficult questions on life and being.

Here is a science after my own heart. Here is the joy and the wonder of living one life in the world. And Hudson's recommendation applies equally well to the study of birds or trees or wildflowers or fossils or fungi, for that matter. Nature is everywhere, costs nothing, requires minimal equipment, tempts us to explore the country and take pleasant walks, is beautiful, and poses difficult questions on life and being. What more could we ask for? Maybe such a science fails to win grants in these days of tight budgets and slavish utility, nor could it pass the litmus test of profit in the marketplace, but a science practiced in this vein would surely preserve more of the world itself and more of our wonderment in it.

And yet, as Hudson surely knew, there's no accounting for wonder when it comes to the animalcules. Some people succumb while others do not. My wife, for instance, finds the animalcules about one-third as mind-boggling as I do, maybe less. She has her own interests, her own sources of wonder, and sometimes grows frustrated when I try to foist my wonder off on her. Recently, for example, a robust colony of rotifers bloomed overnight in one of my samples. Under the scope the colony looked like a ravenous bouquet of two-eyed daisies, each one whirring and trembling with ciliated wheels. (It's hard to describe these crazy things. As one of Leeuwenhoek's draftsmen exclaimed while trying to draw a rotifer for the Royal Society, "O that one could ever depict so wonderful a motion!"). These rotifers were a brand-new species for me, so I called from the study window to my wife, who was out in the garden, telling her that *this* was something she just *had* to see. She paused, looking up from the flowers and patiently removing her gardening gloves, then came inside. Peering at the rotifers, she smiled gently. She did in fact find them rather amusing. But then she stood up and gave me a look of pity. "Honey," she said, her voice tinged with exasperation, "I am very glad that you found a new species of rotifer, and I love that you get such a kick out it, but I am going out now to work in the garden. *Please* don't call me in again."

Some years back I nursed a half-dozen cultures until Christmas and drove them home to my parents' house for a holiday reunion. Childless myself, I wanted to share the wonder with my nieces and nephews. In I came through a front door framed in bright garlands, bearing a box of smelly brown jars like so many putrid gifts to pass around. My sisters looked askance. My teenage nephew laughed at my homemade microscope illuminator — an admittedly preposterous contraption composed of a rotary dimmer switch mounted inside an Irish Cream instant-coffee tin, with a thirty-watt spotlight bulb mounted atop the tin using a leftover bracket originally designed to bolt a toolbox onto a tractor fender (I know, I know: preposterous. But it works surprisingly well).

At first my nieces and nephews were curious about the scope, hefting it and trying all the knobs, but I quickly realized they might have found it more interesting if it came equipped with some kind of laser that could bore holes through whatever hapless victims were trapped on the slide. Nonetheless, I patiently assembled the apparatus for them. And then I started the parade of life itself: I showed them the *Paramecium*, the *Stentor*, the rotifer — these robust and swift creatures that none of them had ever imagined before, much less seen. (I get the impression that students don't spend much time looking at living things in biology classes anymore. What with all the technical urgency of genetic engineering and the remote glamour of the rainforest, there isn't much time to look at real life from something as mundane as a local mud puddle.)

So my younger relations peered down the tube at the spectacles, uttering some "Ewws" and "Wows," and then my youngest niece, after about six or eight seconds, looked up at me and said, "Can we look at something else?"

The quick loss of interest surprised me, because when I was their age I absolutely craved a microscope like this. Instead, my first scope was a cheap toy and a huge disappointment. It came in a brightly painted, hinged metal box and promised to magnify "UP TO 700x!" But in practice it wasn't much better than a magnifying glass, and at higher powers every specimen blurred to a gray murk. I do remember using it one afternoon while sitting at a picnic table on the patio of my boyhood home in the Cincinnati suburbs. On a hot and humid day, I was out there in the sun looking at mosquito larvae, which are plenty big enough to be viewed even with a toy. I watched one twisting larva so long that the water dried up and it slowly died. I remember wondering about its life and wondering some more about its death. I drew pictures of it, speculating about which part was the brain, which the heart, and guessing how big a thought it could think. I felt vaguely guilty for killing it, then I wiped the slide off on my shorts and fished another one from a bucket of stale water. I was at it so long my mother finally came out and asked me what on earth I was doing. When I told her about the larvae, she said that was good, that was fine — but quickly added that I could *not* keep any mosquitoes as pets.

As a lifelong devotee of life, I have had my share of pets. Back then I had a dog, a cat, and some white mice that perpetually escaped. At various times I had tadpoles and toads, frogs and snakes, a box turtle, a painted turtle, and a rabbit. I never had an ant farm — an oversight on my part. In a story she tells nearly every Christmas, my mother swears I once rode my bicycle down to Sheed Creek, a few miles from home, collected thirteen water snakes in a white tube sock and put them all in an aquarium in my bedroom. According to her, by the next morning every last snake had escaped, and she refused to let me leave the house, even for school, until I caught them all again. She claims it took me all morning.

In my version there were only three snakes — a large one and two small ones (I mean, come on: thirteen snakes at once? I should be so lucky). I readily concede the part about the escape, however, noting that I never could figure out how they did it. And regardless of the number of snakes involved, our memories do agree on one beautiful detail: I found the last snake, the smallest, coiled up bright and neat as an enameled brooch in the heel of one of her shoes.

There were plenty of pets, but the oddest must have been *Paramecium caudatum*. A family friend who worked as a research scientist gave me a pure culture of this species, commonly used in lab experiments, and I raised them in beakers, feeding them grains of boiled rice and wheat germ. Actually, bacteria fed on the rice and the *Paramecium* fed on the bacteria. They have prodigious appetites for bacteria, and each one can eat over 5 million per day. Unless I overfed them, they kept their cage scrupulously clean and the water spotless. I suppose the oddest part of the whole episode was that I had no microscope at the time and so couldn't even see them except as the tiniest of specks. Holding a beaker under strong light before a dark background, I found that my young eyes could just make them out: silvery white motes wandering and wobbling through the liquid. With a magnifying glass I could easily discern their peanut shape. I thought they looked like tiny angels hovering in the light. A hundred, at least, could have danced on the head of a pin.

This midsummer afternoon I find only a few skittish *Paramecium*. As odd as it may sound, the pond is too clean for them. They thrive in organically enriched water with an abundance of bacteria, and right now much of the bacteria is all eaten up. Under the microscope they look like brown slippers or peanuts with a groove down one side, and they swim along tracing a helix while sifting out bacteria. They stop to graze in the detritus like cows in a pasture, moving ahead slowly, backing up, moving off again. If bacteria are especially plentiful, they sit motionless and feed like cows at a trough.

They have enemies of course, *Didinium* chief among them. *Didinium* looks like the barrel-shaped housing of an airplane engine, complete with a protruding spinner cone and two rings of cilia zinging around like propellers. It can swallow a *Paramecium* five times its size, but its hunting prowess leaves much to be desired. Mostly *Didinium* swims around at top speed, blindly jabbing its poisoned snout into everything it meets. I've watched them crashing into rotifers and *Euplotes* and *Stentor* and clumps of detritus — even the glass slide. But only *Paramecium* is configured such that the snout will release its quiver of toxic arrows. Imagine a hunter on the African savanna, blindfolded, running around with a poison-tipped spear held out in front. The spear will penetrate only the skin of a zebra and only if it hits head on. Glancing blows don't count. The hunter runs around and around poking elephants, warthogs, rhinos, trees, rocks, the dirt itself — all to no effect. He grazes some zebras, too, but always a glancing blow with a shoulder or a hip, and the zebras scoot safely away. Hour after hour this continues, the hunter bouncing around like a pinball. It's all the hunter can do until, at long last, the improbable happens. The spear hits a zebra broadside, enters easily, and the poison takes effect. The zebra is quickly paralyzed, legs aquiver. Then the hunter does an even more improbable thing: he slowly stretches his mouth over the zebra and swallows it whole.

To decrease the odds of success, *Paramecium* easily outswims *Didinium* in the open, and *Didinium* has trouble negotiating the detritus, which is the only place that *Paramecium* sits still. As one writer succinctly puts it, "*Didiniums* work hard for a living."

Paramecium is probably the best known of all the protozoa. Entire books have been written about its anatomy and physiology, and in 1906 Herbert Spencer Jennings devoted four chapters of his classic *Behavior of the Lower Organisms* to this animalcule. *Paramecium* lives by a fairly simple set of behavioral rules. It advances on bacteria and retreats from nearly everything else. It backs away from obstacles, backs away from acids and salts, backs away from heat and cold and experimental prodding. Every time it backs away, it turns on its heel and heads blindly off in a new direction. This "trial-and-error" behavior is the hallmark of the animalcule lifestyle and has been recognized as the origin of all higher modes of animal behavior.

Although heralded as one of the founders of behaviorism, Jennings reached some surprising conclusions about the animalcules. He began by noting that consciousness must always be inferred in other organisms, since we have no direct access to their experience. He then identified "objective correlates" in the behavior of animalcules for the psychic states such as hunger, sleep, and fear, which we naturally attribute to higher animals. In other words, we can call *Paramecium* "hungry" or "fearful" based on the same behaviors used for inferring those psychic states in higher organisms. If it applies to a dog, Jennings argued, there's no *objective* reason to deny it to the *Paramecium*. In fact, Jennings described the *Paramecium* as a "little subject" — a designation that I like very much. He identified the necessary components of any behavior, including perception, discrimination, attention, and choice. Choice, for instance, means that *Paramecium* must choose between competing stimuli — say, between advancing on bacteria or avoiding the experimenter's drop of acid. He saw that *Paramecium* could make a bad choice in such cases and thus perish. But Jennings was circumspect about this, writing that "choice is not perfect . . . in either lower or higher organisms. Paramecium at times accepts things that are useless or harmful to it, but perhaps on the whole less often than does man."

The most significant conclusion Jennings reached is that protozoan behavior can be conditioned. Responses to stimuli that appear to be reflexes (mere biochemical mechanics) can be modified through repetition and association. *Stentor*, for instance, becomes inured to repeated prodding and ceases to contract defensively. This complicates any suggestion that the behavior of these organisms is merely reflexive, like the reflex of the human leg when tapped below the kneecap. If *Stentor* recognizes nothing else, Jennings argued, at least it recognizes a false alarm.

Jennings reached his most controversial conclusions in the final chapters, where he writes:

> The writer is thoroughly convinced, after long study of the behavior of this organism, that if Amoeba were a large animal, so as to come within the everyday experience of human beings, its behavior would at once call forth the attribution to it of states of pleasure and pain, of hunger, desire, and the like, on precisely the same basis as we attribute these things to the dog . . . In conducting objective investigations we train ourselves to suppress this impression, but thorough investigation tends to restore it stronger than at first.

Of course Jennings is not arguing here that Amoeba *was* conscious, only that it passed the behavioral tests we rely on to infer consciousness in higher organisms, including other humans. If we give a dog the benefit of the doubt when it comes to consciousness, then objectively we might give the same to Amoeba.

When I think of this I'm not so sure. I am fond of life in all its forms and give it all the credit I can, but these animalcules can be unnerving. Sometimes while watching them go through their sleepless routine a little sea change comes over me. One moment I see Jennings's "little subjects," the ghosts piloting their various machines, and the next moment I catch a glimpse of the machine behind the ghost, and a suspicion arises that deep down no ghosts are required, no strings need to be pulled. The molecules are the only strings, and they pull themselves. Maybe *Paramecium* requires fifty chalkboards of biochemical equations all humming in perfect equilibrium to maintain itself and qualify as life. The question remains: what more is required? Where would it come from?

Most scientists, of course, no longer talk about Jennings's "little subjects." They describe such language as anthropomorphism and avoid it like the plague. But this avoidance overlooks some of the toughest and most interesting questions. Perhaps without intending to, many scientists are steadily demoting all life, both high and low, to the level of machinery. In a preface to a recent edition of Jennings's book, I find that instead of observing animalcule behavior, scientists are now "occupied with finding and fitting together the nuts and bolts of the machinery in cellular architecture, metabolism, heredity, development, and, now, motility and behavior."

I won't argue with that, except to wonder about where in the development of animal behavior the "little subjects" begin piloting "the nuts and bolts of the machinery." Perhaps they never really do. Maybe we are all machines, the end products of 4 billion years of molecular watchmaking. Maybe consciousness is only a useful fiction that yields an evolutionary advantage, and the "I" I endlessly refer to is nothing but a slippery epiphenomenon, the byproduct of a highly wrought and well-adapted nervous system. Along these lines the Harvard psychologist Steven Pinker recently declared that "the soul is the activity of the brain." This sounds fairly definitive, but the moment you think about it you realize it doesn't really explain much. It's a lot like saying van Gogh's *Starry Night* is 5,637 brushstrokes applied to a canvas measuring 74 x 92 cm.

In order to escape this reductive conundrum, others have suggested that mind is an emergent property, one that cannot be reduced to or predicated upon the firing of neurons and the levels of dopamine. Instead it is a new level of organization, more than the sum of its parts, in the same way, perhaps, that ecosystems and Gaia are said to be. The challenge here is to deflect charges that you have merely filled a gap in the mechanism with a leap of faith. As yet, of course, the question of consciousness and mind remains unanswered. But one thing is certain, at least from where I sit: consciousness is one hell of an interesting ride till death do us part.

I could go on through this world in the pickle jar, placing drop after drop on the stage and finding one bizarre life form after another. More than 30,000 species of protists have been described so far, but of course there are many more than that, with estimates ranging up to 200,000. A man named L. R. Cleveland spent forty years studying the wood roach *Cryptocercus punctulatus*, a distant and uncommon relative of our domestic variety. He devoted an entire volume to this insect, in which he describes twenty-six species of flagellates living in the hindgut alone, strange little organisms often veiled in streams of flagella like long, flowing hair. Symbionts mostly, these flagellates digest the wood cellulose that the roach could not otherwise digest. Others no doubt prey on the symbionts, while still others probably parasitize the roach. Not even Cleveland had the time to sort it all out. The point is, years later another scientist revisited this bug, spread some fresh hindgut on a microscope slide, and described thirty-one brand-new species that Cleveland missed the first time around. That's fifty-seven species and counting, all living happily in a single bug's gut.

The first axiom of naturalists everywhere must surely be "the more you look, the more you find." I have identified a few of the animalcules in this jar, but I confess most of them remain mysteries to me. I quickly sketch a tiny flagellate and move on. A streamlined ciliate zips across the field, and I let it pass. And beyond this realm of Protista lies the deeper realm of bacteria, beyond that the nanobacteria, then the viruses, the viroids, the prions — entities becoming ever smaller and simpler until life as we know it, and somewhat arbitrarily define it, eases smoothly off into nonlife, into globules and membranes and strings of organic molecules assembling spontaneously in the ooze. At this level, stumped scientists often resort to the term "biological particle" instead of "life."

If you ask someone who has spent the afternoon with a jar of pond water and a microscope to sum up the life of the animalcules in a single word, the reply would surely be "eating." Eating and eating and eating. It's eating all the way down, you might say. There are, of course, some autotrophs — the algae, the dinoflagellates, the euglenoids — all of whom can produce food using chlorophyll the same way plants do, but all the rest will be found busily, never-endingly eating. Beyond that, the only requirement is reproduction, and here the efficient pace continues. Most of these organisms skip the romance, skip sex and gender altogether, and head straight to mitosis. They make a copy of their genes, then slowly split in half (eating all the while, of course). In the final moments, just before the cells pinch apart, they tug and strain against each other, each seemingly eager to be rid of the other. They are perfect twins, and yet there's no recognition of any kind, no legacy beyond the coiled chain of the gene. Having witnessed the relentless pace, the precariousness, the continuous offhand waste of animalcule life, I consider myself very lucky to be way up here atop the food chain sipping iced tea on my porch in the afternoon. We truly are pinnacles, though mostly we forget this and don't behave like it.

Know Thyself — the ancient call to wisdom. You should know, then, that what works inside this jar works just as well inside of you. The strategies and structures are all the same. All of life springs tried and true, so that, at the cellular level, we ape the protozoans no end. Ciliated cells line passages throughout our bodies. They line the respiratory system and filter dust from the air we breath. We smell with cilia, hundreds of millions of them densely packed. Cilia line the fallopian tubes to whisk an egg from the ovary down to the uterus. Along the way, this ovum might meet a counterforce in the form of a flagellated sperm cell working tirelessly against the cilia's currents. Amoebas, too — yet another idea too good to leave behind. Each cubic millimeter of our blood contains thousands of them, the white blood cells, roving through every passage in the body and engulfing intruders in pseudopods. We humans are a complex but ultimately conservative assortment of the tried and true. I can't help but smile when I think of this. I feel like a colonial organism. A wildly improbable group effort. What are my amoebas to me? And what am I to them? We are symbionts in a way, partners living together for the common good. I provide a home and hearth, while they do all the cleaning.

But look closer: even the fire in my hearth may not be my own. It is supplied instead by the mitochondria, organelles often referred to as the "power plants" of the cell, and which, properly speaking, are not quite human at all, since they retain their own genes, their own membranes, and originally had their own reasons for living within our cells. The general consensus is that they descend from ancient predatory bacteria that invaded other bacteria, eventually settling into a symbiotic living arrangement. Many theorists now believe that most of the structures within complex eukaryotic cells (nucleated plant and animal cells, including all the protists) developed through ongoing endosymbiotic unions between existing life and invading bacteria. This includes structures such as the aforementioned cilia and flagella that may have derived from spirochete (corkscrew) bacteria. One theorist, Lyn Margulis, has gone further still, suggesting that spirochetes may ultimately have formed the basis of the animal brain and consciousness itself.

Which reminds me: about this time a year ago I was dueling with a tenacious little spirochete named *Borrelia burgdorferi*, a bacterium better known as Lyme disease. I had suspected I might have the disease for about a month and was closely tracking the symptoms until I had enough cause to visit the doctor; then one morning I woke up with Bell's palsy, which is one-sided facial paralysis and a real classic sign of the disease. I looked in the mirror while shaving but managed only a crooked half smile. I swiped at a bit of drool that ran down through the shaving cream. That's it, I thought. That's all the evidence *I* need. I was off to the doctor in a hurry.

During a month-long course of potent antibiotics, I thought often about this bacterium that was traveling through my blood and lymph, gathering here in my neck between the shoulder blades, there at a knee joint, once more in a facial nerve, all the while reproducing in the fast and elemental way of all bacteria. While reading one day out here on the porch, I caught myself just staring at the back of my hand while slowly flexing the tendons and knuckle bones. Memories of wriggling spirochetes I had seen under the scope were now projected under the skin and deep into the joints of me. I was a little confused, because somehow the presence of *Borrelia* did not seem particularly unnatural to me. Not exactly at home, maybe, but neither out of place. I knew there were equally improbable things at work within me, such as osteoclasts and osteoblasts. These are highly specialized bone cells whose jobs are tunneling through living bone, honeycombing it and redepositing it along changing lines of skeletal stress. Their combined efforts allow something as hard as bone to be reworked, revised, and healed through time, allow it to grow and change and adapt. *Borrelia* seemed only a bit stranger than that, only *slightly* less at home, just a *little* more foreign. That *Borrelia* could do me more harm than good seemed almost the sole criterion for judgment in the matter. If my bones are riddled by osteoclasts it turns out well, if riddled by *Borrelia* it turns out badly.

I was, of course, grateful for the antibiotics that wiped out *Borrelia* within the month. Without that medicine I might not be writing these words now. I might be crippled; I might be dead. But the thing itself, the bacteria — "those indefatigable little junkmen" Donald Culross Peattie called them — somehow, they just didn't seem out of place. At the time I wondered if these thoughts were wisdom or lunacy. I still do.

Who really knows? All this intricacy and all this craziness just to keep me walking around in one piece. It's late afternoon now, and I return, blissfully intact, to the shore of the pond with the world in the pickle jar sloshing over the rim. I hold the jar up to the evening light and gaze through it a final time. Look close and see the angels dancing. I have identified more than forty species of life in there today, but I'm only a novice with a spare afternoon. Perhaps one hundred is a better guess. And two hundred wouldn't surprise me at all. The truth is I have no idea. Life fills and overfills the world from puddle to ocean, from dirt clod to mountaintop. When I pour the living water into the pond, it gleams for an instant. Then the pond takes it, absorbs it seamlessly, smoothes the ripples over again.

MICHAEL SPECTER

Big Foot

FROM *The New Yorker*

A LITTLE MORE than a year ago, Sir Terry Leahy, who is the chief executive of the Tesco chain of supermarkets, Britain's largest retailer, delivered a speech to a group called the Forum for the Future, about the implications of climate change. Leahy had never before addressed the issue in public, but his remarks left little doubt that he recognized the magnitude of the problem. "I am not a scientist," he said. "But I listen when the scientists say that if we fail to mitigate climate change, the environmental, social, and economic consequences will be stark and severe . . . There comes a moment when it is clear what you must do. I am determined that Tesco should be a leader in helping to create a low-carbon economy. In saying this, I do not underestimate the task. It is to take an economy where human comfort, activity, and growth are inextricably linked with emitting carbon and to transform it into one which can only thrive without depending on carbon. This is a monumental challenge. It requires a revolution in technology and a revolution in thinking. We are going to have to rethink the way we live and work."

Tesco sells nearly a quarter of the groceries bought in the United Kingdom, it possesses a growing share of the markets in Asia and Europe, and late last year the chain opened its first stores in America. Few corporations could have a more visible — or forceful — impact on the lives of their customers. In his speech, Leahy, who is fifty-two, laid out a series of measures that he hoped would ignite "a revolution in green consumption." He announced that Tesco would cut its energy use in half by 2010, drastically limit the number of products it transports by air, and place airplane symbols on the packaging of those which it does. More important, in an effort to help consumers understand the environmental impact of the choices they make every day, he told the forum that Tesco would develop a system of carbon labels and put them on each of its 70,000 products. "Customers want us to develop ways to take complicated carbon calculations and present them simply," he said. "We will therefore begin the search for a universally accepted and commonly understood measure of the carbon footprint of every product we sell — looking at its complete life cycle, from production through distribution to consumption. It will enable us to label all our products so that customers can compare their carbon footprint as easily as they can currently compare their price or their nutritional profile."

Leahy's sincerity was evident, but so was his need to placate his customers. Studies have consistently demonstrated that, given a choice, people prefer to buy products that are environmentally benign. That choice, however, is almost never easy. "A carbon label will put the power in the hands of consumers to choose how they want to be green," Tom Delay, the head of the British government's Carbon Trust, said. "It will empower us all to make informed choices and in turn drive a market for low-carbon products." Tesco was not alone in telling people what it would do to address the collective burden of our greenhouse-gas emissions. Compelled by economic necessity as much as by ecological awareness, many corporations now seem to compete as vigorously to display their environmental credentials as they do to sell their products.

In Britain, Marks & Spencer has set a goal of recycling all its waste, and intends to become carbon-neutral by 2012 — the equivalent, it claims, of taking a hundred thousand cars off the road every year. Kraft Foods recently began to power part of a New York plant with methane produced by adding bacteria to whey, a by-product of cream cheese. Not to be outdone, Sara Lee will deploy solar panels to run one of its bakeries in New Mexico. Many airlines now sell "offsets," which offer passengers a way to invest in projects that reduce CO_2 emissions. In theory, that would compensate for the greenhouse gas caused by their flights. This year's Super Bowl was fueled by wind turbines. There are carbon-neutral investment banks, carbon-neutral real-estate brokerages, carbon-neutral taxi fleets, and carbon-neutral dental practices. Detroit, arguably America's most vivid symbol of environmental excess, has also staked its claim. ("Our designers know green is the new black," Ford declares on its home page. General Motors makes available hundreds of green pictures, green stories, and green videos to anyone who wants them.)

Possessing an excessive carbon footprint is rapidly becoming the modern equivalent of wearing a scarlet letter. Because neither the goals nor acceptable emissions limits are clear, however, morality is often mistaken for science. A recent article in *New Scientist* suggested that the biggest problem arising from the epidemic of obesity is the additional carbon burden that fat people — who tend to eat a lot of meat and travel mostly in cars — place on the environment. Australia briefly debated imposing a carbon tax on families with more than two children; the environmental benefits of abortion have been discussed widely (and simplistically). Bishops of the Church of England have just launched a "carbon fast," suggesting that during Lent parishioners, rather than giving up chocolate, forgo carbon. (Britons generate an average of a little less than ten tons of carbon per person each year; in the United States, the number is about twice that.)

Greenhouse-gas emissions have risen rapidly in the past two centuries, and levels today are higher than at any time in at least the past 650 thousand years. In 1995, each of the 6 billion people on Earth was responsible, on average, for one ton of carbon emissions. Oceans and forests can absorb about half that amount. Although specific estimates vary, scientists and policy officials increasingly agree that allowing emissions to continue at the current rate would induce dramatic changes in the global climate system. To avoid the most catastrophic effects of those changes, we will have to hold emissions steady in the next decade, then reduce them by at least 60 to 80 percent by the middle of the century. (A delay of just ten years in stopping the increase would require double the reductions.) Yet even if all carbon emissions stopped today, the Earth would continue to warm for at least another century. Facts like these have transformed carbon dioxide into a strange but powerful new currency, difficult to evaluate yet impossible to ignore.

A person's carbon footprint is simply a measure of his contribution to global warming. (CO_2 is the best known of the gases that trap heat in the atmosphere, but others — including water vapor, methane, and nitrous oxide — also play a role.) Virtually every human activity — from watching television to buying a quart of milk — has some carbon cost associated with it. We all consume electricity generated by burning fossil fuels; most people rely on petroleum for transportation and heat. Emissions from those activities are not hard to quantify. Watching a plasma television for three hours every day contributes 250 kilograms of carbon to the atmosphere each year; an LCD television is responsible for less than half that number. Yet the calculations required to assess the full environmental impact of how we live can be dazzlingly complex. To sum them up on a label will not be easy. Should the carbon label on a jar of peanut butter include the emissions caused by the fertilizer, calcium, and potassium applied to the original crop of peanuts? What about the energy used to boil the peanuts once they have been harvested or to mold the jar and print the labels? Seen this way, carbon costs multiply rapidly. A few months ago, scientists at the Stockholm Environment Institute reported that the carbon footprint of Christmas — including food, travel, lighting, and gifts — was 650 kilograms per person. That is as much, they estimated, as the weight of "one thousand Christmas puddings" for every resident of England.

As a source of global warming, the food we eat — and how we eat it — is no more significant than the way we make clothes or travel or heat our homes and offices. It certainly doesn't compare to the impact made by tens of thousands of factories scattered throughout the world. Yet food carries enormous symbolic power, so the concept of "food miles" — the distance a product travels from the farm to your home — is often used as a kind of shorthand to talk about climate change in general. "We have to remember our goal: reduce emissions of greenhouse gases," John Murlis told me not long ago when we met in London. "That should be the world's biggest priority." Murlis is the chief scientific adviser to the Carbon Neutral Company, which helps corporations adopt policies to reduce their carbon footprint as well as those of the products they sell. He has also served as the director of strategy and chief scientist for Britain's Environment Agency. Murlis worries that in our collective rush to make choices that display personal virtue, we may be losing sight of the larger problem. "Would a carbon label on every product help us?" he asked. "I wonder. You can feel very good about the organic potatoes you buy from a farm near your home, but half the emissions — and half the footprint — from those potatoes could come from the energy you use to cook them. If you leave the lid off, boil them at a high heat, and then mash your potatoes, from a carbon standpoint you might as well drive to McDonald's and spend your money buying an order of French fries."

One particularly gray morning last December, I visited a Tesco store on Warwick Way, in the Pimlico section of London. Several food companies have promised to label their products with the amount of carbon-dioxide emissions associated with making and transporting them. Last spring Walkers crisps (potato chips) became the first of them to reach British stores, and they are still the only product on the shelves there with a carbon label. I walked over to the crisp aisle, where a young couple had just tossed three bags of Walkers Prawn Cocktail crisps into their shopping cart. The man was wearing fashionable jeans and sneakers without laces. His wife was toting a huge Armani Exchange bag on one arm and dragging their four-year-old daughter with the other. I asked if they paid attention to labels. "Of course," the man said, looking a bit insulted. He was aware that Walkers had placed a carbon label on the back of its crisp packages; he thought it was a good idea. He just wasn't sure what to make of the information.

Few people are. In order to develop the label for Walkers, researchers had to calculate the amount of energy required to plant seeds for the ingredients (sunflower oil and potatoes), as well as to make the fertilizers and pesticides used on those potatoes. Next, they factored in the energy required for diesel tractors to collect the potatoes, then the effects of chopping, cleaning, storing, and bagging them. The packaging and printing processes also emit carbon dioxide and other greenhouse gases, as does the petroleum used to deliver those crisps to stores. Finally, the research team assessed the impact of throwing the empty bags in the trash, collecting the garbage in a truck, driving to a landfill, and burying them. In the end, the researchers — from the Carbon Trust — found that 75 grams of greenhouse gases are expended in the production of every individual-size bag of potato chips.

"Crisps are easy," Murlis had told me. "They have only one important ingredient, and the potatoes are often harvested near the factory." We were sitting in a deserted hotel lounge in central London, and Murlis stirred his tea slowly, then frowned. "Let's just assume every mother cares about the environment — what then?" he asked. "Should the carbon content matter more to her than the fat content or the calories in the products she buys?"

I put that question to the next shopper who walked by, Chantal Levi, a Frenchwoman who has lived in London for thirty-two years. I watched her grab a large bag of Doritos and then, shaking her head, return it to the shelf. "Too many carbohydrates," she said. "I try to watch that, but between the carbs and the fat and the protein it can get to be a bit complicated. I try to buy locally grown, organic food," she continued. "It tastes better, and it's far less harmful to the environment." I asked if she was willing to pay more for products that carried carbon labels. "Of course," she said. "I care about that. I don't want my food flown across the world when I can get it close to home. What a waste."

It is a logical and widely held assumption that the ecological impacts of transporting food — particularly on airplanes over great distances — are far more significant than if that food were grown locally. There are countless books, articles, Web sites, and organizations that promote the idea. There is even a "100-Mile Diet," which encourages participants to think about "local eating for global change." Eating locally produced food has become such a phenomenon, in fact, that the word "locavore" was just named the 2007 word of the year by the New Oxford American Dictionary.

Paying attention to the emissions associated with what we eat makes obvious sense. It is certainly hard to justify importing bottled water from France, Finland, or Fiji to a place like New York, which has perhaps the cleanest tap water of any major American city. Yet according to one recent study, factories throughout the world are burning 18 million barrels of oil and consuming 41 billion gallons of fresh water every day, solely to make bottled water that most people in the United States don't need.

"Have a quick rifle through your cupboards and fridge and jot down a note of the countries of origin for each food product," Mark Lynas wrote in his popular handbook *Carbon Counter*, published last year by HarperCollins. "The further the distance it has traveled, the bigger the carbon penalty. Each glass of orange juice, for example, contains the equivalent of two glasses of petrol once the transport costs are included. Worse still are highly perishable fresh foods that have been flown in from far away — green beans from Kenya or lettuce from the U.S. They may be worth several times their weight in jet fuel once the transport costs are factored in."

Agricultural researchers at the University of Iowa have reported that the food miles attached to items that one buys in a grocery store are twenty-seven times higher than those for goods bought from local sources. American produce travels an average of nearly 1,500 miles before we eat it. Roughly 40 percent of our fruit comes from overseas, and even though broccoli is a vigorous plant grown throughout the country, the broccoli we buy in a supermarket is likely to have been shipped 1,800 miles in a refrigerated truck. Although there are vast herds of cattle in the United States, we import 10 percent of our red meat, often from as far away as Australia or New Zealand.

In his speech last year, Sir Terry Leahy promised to limit to less than 1 percent the products that Tesco imports by air. In the United States, many similar efforts are underway. Yet the relationship between food miles and their carbon footprint is not nearly as clear as it might seem. That is often true even when the environmental impact of shipping goods by air is taken into consideration. "People should stop talking about food miles," Adrian Williams told me. "It's a foolish concept: provincial, damaging, and simplistic." Williams is an agricultural researcher in the Natural Resources Department of Cranfield University in England. He has been commissioned by the British government to analyze the relative environmental impacts of a number of foods. "The idea that a product travels a certain distance and is therefore worse than one you raised nearby — well, it's just idiotic," he said. "It doesn't take into consideration the land use, the type of transportation, the weather, or even the season. Potatoes you buy in winter, of course, have a far higher environmental ticket than if you were to buy them in August." Williams pointed out that when people talk about global warming they usually speak only about carbon dioxide. Making milk or meat contributes less CO_2 to the atmosphere than building a house or making a washing machine. But the animals produce methane and nitrous oxide, and those are greenhouse gases, too. "This is not an equation like the number of calories or even the cost of a product," he said. "There is no one number that works."

Many factors influence the carbon footprint of a product: water use, cultivation and harvesting methods, quantity and type of fertilizer, even the type of fuel used to make the package. Sea-freight emissions are less than a sixtieth of those associated with airplanes, and you don't have to build highways to berth a ship. Last year a study of the carbon cost of the global wine trade found that it is actually more "green" for New Yorkers to drink wine from Bordeaux, which is shipped by sea, than wine from California, sent by truck. That is largely because shipping wine is mostly shipping glass. The study found that "the efficiencies of shipping drive a 'green line' all the way to Columbus, Ohio, the point where a wine from Bordeaux and Napa has the same carbon intensity."

The environmental burden imposed by importing apples from New Zealand to northern Europe or New York can be lower than if the apples were raised fifty miles away. "In New Zealand, they have more sunshine than in the UK, which helps productivity," Williams explained. That means the yield of New Zealand apples far exceeds the yield of those grown in northern climates, so the energy required for farmers to grow the crop is correspondingly lower. It also helps that the electricity in New Zealand is mostly generated by renewable sources, none of which emit large amounts of CO_2. Researchers at Lincoln University in Christchurch found that lamb raised in New Zealand and shipped 11,000 miles by boat to England produced 688 kilograms of carbon-dioxide emissions per ton, about a fourth the amount produced by British lamb. In part that is because pastures in New Zealand need far less fertilizer than most grazing land in Britain (or in many parts of the United States). Similarly, importing beans from Uganda or Kenya — where the farms are small, tractor use is limited, and the fertilizer is almost always manure — tends to be more efficient than growing beans in Europe, with its reliance on energy-dependent irrigation systems.

Williams and his colleagues recently completed a study that examined the environmental costs of buying roses shipped to England from Holland and of those exported (and sent by air) from Kenya. In each case, the team made a complete life-cycle analysis of 12,000 rose stems for sale in February — in which all the variables, from seeds to store, were taken into consideration. They even multiplied the CO_2 emissions for the air-freighted Kenyan roses by a factor of nearly three to account for the increased effect of burning fuel at a high altitude. Nonetheless, the carbon footprint of the roses from Holland — which are almost always grown in a heated greenhouse — was six times the footprint of those shipped from Kenya. Even Williams was surprised by the magnitude of the difference. "Everyone always wants to make ethical choices about the food they eat and the things they buy," he told me. "And they should. It's just that what seems obvious often is not. And we need to make sure people understand that before they make decisions on how they ought to live."

How do we alter human behavior significantly enough to limit global warming? Personal choices, no matter how virtuous, cannot do enough. It will also take laws and money. For decades American utilities built tall smokestacks, hoping to keep the pollutants they emitted away from people who lived nearby. As emissions are forced into the atmosphere, however, they react with water molecules and then are often blown great distances by prevailing winds, which in the United States tend to move from west to east. Those emissions — principally sulfur dioxide produced by coal-burning power plants — are the primary source of acid rain, and by the 1970s it had become clear that they were causing grave damage to the environment and to the health of many Americans. Adirondack Park, in upstate New York, suffered more than anywhere else: hundreds of streams, ponds, and lakes there became so acidic that they could no longer support plant life or fish. Members of Congress tried repeatedly to introduce legislation to reduce sulfur-dioxide levels, but the Reagan administration (as well as many elected officials, both Democratic and Republican, from regions where sulfur-rich coal is mined) opposed any controls, fearing that they would harm the economy. When the cost of polluting is negligible, so are the incentives to reducing emissions.

"We had a complete disaster on our hands," Richard Sandor told me recently, when I met with him at his office at the Chicago Climate Exchange. Sandor, a dapper sixty-six-year-old man in a tan cable-knit cardigan and round, horn-rimmed glasses, is the exchange's chairman and CEO. In most respects, the exchange operates like any other market. Instead of pork-belly futures or gold, however, CCX members buy and sell the right to pollute. Each makes a voluntary (but legally binding) commitment to reduce emissions of greenhouse gases — including carbon dioxide, methane, and nitrous oxide—and hydrofluorocarbons. Four hundred corporations now belong to the exchange, including a growing percentage of America's largest manufacturers. The members agree to reduce their emissions by a certain amount every year, a system commonly known as cap and trade. A baseline target, or cap, is established, and companies whose emissions fall below that cap receive allowances, which they can sell (or save to use later). Companies whose emissions exceed the limit are essentially fined and forced to buy credits to compensate for their excess.

Sandor led me to the "trading floor," which, like most others these days, is a virtual market populated solely by computers. "John, can you get the carbon futures up on the big screen?" Sandor yelled to one of his colleagues. Suddenly, a string of blue numbers slid across the monitor. "There is our 2008 price," Sandor said. Somebody had just bid $2.15 per ton for carbon futures.

A former Berkeley economics professor and chief economist at the Chicago Board of Trade, Sandor is known as the "father of financial futures." In the 1970s, he devised a market in interest rates which, when they started to fluctuate, turned into an immense source of previously untapped wealth. His office is just north of the Board of Trade, where he served for two years as vice-chairman. The walls are filled with interest-rate arcana and mortgage memorabilia; his desk is surrounded by monitors that permit him to track everything from catastrophic-risk portfolios to the price of pollution.

Sandor invents markets to create value for investors where none existed before. He sees himself as "a guy from the sixties" — but one who believes that free markets can make inequality disappear. So, he wondered, why not offer people the right to buy and sell shares in the value of reduced emissions? "At first, people laughed when I suggested the whole future idea," he said. "They didn't see the point of hedging on something like interest rates, and when it came to pollution rights many people just thought it was wrong to take a business approach to environmental protection."

For Sandor, personal factors like food choices and driving habits are small facets of a far larger issue: making pollution so costly that our only rational choice is to stop. When he started, though, the idea behind a sulfur-dioxide-emissions market was radical. It also seemed distasteful; opponents argued that codifying the right to pollute would only remove the stigma from an unacceptable activity. You can't trade something unless you own it; to grant a company the right to trade in emissions is also to give it a property right over the atmosphere. (This effect was noted most prominently when the Reagan administration deregulated airport landing rights in 1986. Airlines that already owned the rights to land got to keep those rights, while others had to buy slots at auction; in many cases, that meant that the country's richest airlines were presented with gifts worth millions of dollars.)

Sandor acknowledges the potential for abuse, but he remains convinced that emissions will never fall unless there is a price tag attached to them. "You are really faced with a couple of possibilities when you want to control something," he told me. "You can say, 'Hey, we will allow you to use only *x* amount of these pollutants.' That is the command approach. Or you can make a market."

In the late 1980s, Sandor was asked by an Ohio public-interest group if he thought it would be possible to turn air into a commodity. He wrote an essay advocating the creation of an exchange for sulfur-dioxide emissions. The idea attracted a surprising number of environmentalists, because it called for large and specific reductions; conservatives who usually oppose regulation approved of the market-driven solution.

When Congress passed the Clean Air Act in 1990, the law included a section that mandated annual acid-rain reductions of 10 million tons below 1980 levels. Each large smokestack was fitted with a device to measure sulfur-dioxide emissions. As a way to help meet the goals, the act enabled the creation of the market. "Industry lobbyists said it would cost ten billion dollars in electricity increases a year. It cost one billion," Sandor told me. It soon became less expensive to reduce emissions than it was to pollute. Consequently, companies throughout the country suddenly discovered the value of investing millions of dollars in scrubbers, which capture and sequester sulfur dioxide before it can reach the atmosphere.

Sandor still enjoys describing his first sulfur trade. Representatives of a small Midwestern town were seeking a loan to build a scrubber. "They were prepared to borrow millions of dollars and leverage the city to do it," he told me. "We said, 'We have a better idea.'" Sandor arranged to have the scrubber installed with no initial cost, and the apparatus helped the city fall rapidly below its required emissions cap. He then calculated the price of thirty years' worth of that municipality's SO_2 emissions and helped arrange a loan for the town. "We gave it to them at a significantly lower rate than any bank would have done," Sandor said. "It was a fifty-million-dollar deal and they saved seven hundred and fifty thousand dollars a year — and never had to pay a balloon mortgage at the end. I mention this because trading that way not only allows you to comply with the law, but it provides creative financing tools to help structure the way investments are made. It encourages people to comply at lower costs, because then they will make money."

The program has been an undisputed success. Medical and environmental savings associated with reduced levels of lung disease and other conditions have been enormous — more than $100 billion a year, according to the EPA. "When is the last time you heard somebody even talking about acid rain?" Sandor asked. "It was going to ravage the world. Now it is not even mentioned in the popular press. We have reduced emissions from eighteen million tons to nine million, and we are going to halve it again by 2010. That is as good a social policy as you are ever likely to see."

No effort to control greenhouse-gas emissions or to lower the carbon footprint — of an individual, a nation, or even the planet — can succeed unless those emissions are priced properly. There are several ways to do that: they can be taxed heavily, like cigarettes, or regulated, which is the way many countries have established mileage-per-gallon standards for automobiles. Cap and trade is another major approach — although CO_2 emissions are a far more significant problem for the world than those that cause acid rain, and any genuine solution will have to be global.

Higher prices make conservation appealing — and help spark investment in clean technologies. When it costs money to use carbon, people begin to seek profits from selling fuel-efficient products like long-lasting light bulbs, appliances that save energy, hybrid cars, even factories powered by the sun. One need only look at the passage of the Clean Water Act in 1972 to see that a strategy that combines legal limits with realistic pricing can succeed. Water had always essentially been free in America, and when something is free people don't value it. The act established penalties that made it expensive for factories to continue to pollute water. Industry responded at once, and today the United States (and much of the developed world) manufactures more products with less water than it did fifty years ago. Still, whether you buy a plane ticket, an overcoat, a Happy Meal, a bottle of wine imported from Argentina, or a gallon of gasoline, the value of the carbon used to make those products is not reflected by their prices.

In 2006 Sir Nicholas Stern, a former chief economist of the World Bank, who is now the head of Britain's Economic Service, issued a comprehensive analysis of the implications of global warming in which he famously referred to climate change as "the greatest market failure the world has ever seen." Sir Nicholas suggested that the carbon emissions embedded in almost every product ought, if priced realistically, to cost about eighty dollars a ton.

Trading schemes have many opponents, some of whom suggest that attaching an acceptable price to carbon will open the door to a new form of colonialism. After all, since 1850, North America and Europe have accounted for 70 percent of all greenhouse gas emissions, a trend that is not improving. Stephen Pacala, the director of Princeton University's Environmental Institute, recently estimated that half of the world's carbon-dioxide emissions come from just 700 million people, about 10 percent of the population.

If prices were the same for everyone, however, rich countries could adapt more easily than countries in the developing world. "This market driven mechanism subjects the planet's atmosphere to the legal emission of greenhouse gases," the anthropologist Heidi Bachram has written. "The arrangement parcels up the atmosphere and establishes the routinized buying and selling of 'permits to pollute' as though they were like any other international commodity." She and others have concluded that such an approach would be a recipe for social injustice.

No one I spoke to for this story believes that climate change can be successfully addressed solely by creating a market. Most agreed that many approaches — legal, technological, and financial — will be necessary to lower our carbon emissions by at least 60 percent

over the next fifty years. "We will have to do it all and more," Simon
Thomas told me. He is the chief executive officer of Trucost, a consulting firm that helps gauge the full burden of greenhouse-gas
emissions and advises clients on how to address them. Thomas
takes a utilitarian approach to the problem, attempting to con-
vince corporations, pension funds, and other investors that the
price of continuing to ignore the impact of greenhouse-gas emissions will soon greatly exceed the cost of reducing them.

Thomas thinks that people finally are beginning to get the message. Apple computers certainly has. Two years ago, Greenpeace
began a "Green my Apple" campaign, attacking the company for
its "iWaste." Then, in the spring of 2007, not long before Apple
launched the iPhone, Greenpeace issued a guide to electronics
that ranked major corporations on their tracking, reporting, and
reduction of toxic chemicals and electronic waste. Apple came in
last. The group's findings were widely reported, and stockhold-
ers took notice. (A company that sells itself as one of America's
most innovative brands cannot afford to ignore the environmental
consequences of its manufacturing processes.) Within a month,
Steve Jobs, the company's CEO, posted a letter on the Apple Web
site promising a "greener Apple." He committed the company to
ending the use of arsenic and mercury in monitors and said that
the company would shift rapidly to more environmentally friendly
LCD displays.

"The success of approaches such as ours relies on the idea that
even if polluters are not paying properly now there is some reasonable prospect that they will have to pay in the future," Thomas told
me. "If that is true, then we know the likely costs and they are of significant value. If polluters never have to pay, then our approach
will fail.

"You have to make it happen, though," he went on. "And that is
the job of government. It has to set a level playing field so that
a market economy can deliver what it's capable of delivering."
Thomas, a former investment banker, started Trucost nearly a decade ago. He mentioned the free-market economist Friedrich von
Hayek, who won the Nobel Prize in Economics in 1974. "There is a
remarkable essay in which he shows how an explosion, say, in a
South American tin mine could work its way through the global
supply chain to increase the price of canned goods in Europe,"

Thomas said. I wondered what the price of tin could have to do with the cost of global warming.

"It is very much to the point," Thomas answered. "Tin became more expensive and the market responded. In London people bought fewer canned goods. The information traveled all the way from that mine across the world without any person in that supply chain even knowing the reasons for the increase. But there was less tin available, and the market responded as you would have hoped it would." To Thomas the message was simple: "If something is priced accurately, its value will soon be reflected in every area of the economy."

Without legislation, it is hard to imagine that a pricing plan could succeed. (The next administration is far more likely to act than the Bush administration has been. The best-known climate-change bill now before Congress, which would mandate capping carbon limits, was written by Senator Joseph Lieberman. Hillary Clinton, Barack Obama, and John McCain are cosponsors. Most industrial leaders, whatever their ideological reservations, would prefer a national scheme to a system of rules that vary from state to state.) Even at today's anemic rates, however, the market has begun to function. "We have a price of carbon that ranges from two to five dollars a ton," Sandor told me. "And everyone says that is too cheap. Of course, they are right. But it's not too cheap for people to make money.

"I got a call from a scientist a while ago" — Isaac Berzin, a researcher at MIT. "He said, 'Richard, I have a process where I can put an algae farm next to a power plant. I throw some algae in, and it becomes a super photosynthesis machine and sucks the carbon dioxide out of the air like a sponge. Then I gather the algae, dry it out, and use it as renewable energy." Berzin asked Sandor whether, if he was able to take 50 million tons of carbon dioxide out of the atmosphere in this way, he could make $100 million.

"I said, 'Sure,'" Sandor recalled, laughing. "Two dollars a ton, why not? So he sends me a term paper. Not a prospectus, even." Sandor was skeptical, but it didn't take Berzin long to raise $20 million from investors, and he is now working with the Arizona Public Service utility to turn the algae into fuel. Sandor shook his head. "This is at two dollars a ton," he said. "The lesson is important: price stimulates inventive activity. Even if you think the price is too low or ridiculous. Carbon has to be rationed, like water and clean air. But I absolutely promise that if you design a law and a trading scheme properly, you are going to find everyone, from professors at MIT to the guys in Silicon Valley, coming out of the woodwork. That is what we need, and we need it now."

In 1977 Jimmy Carter told the American people that they would have to balance the nation's demand for energy with its "rapidly shrinking resources" or the result "may be a national catastrophe." It was a problem, the president said, "that we will not solve in the next few years, and it is likely to get progressively worse through the rest of this century. We must not be selfish or timid if we hope to have a decent world for our children and grandchildren." Carter referred to the difficult effort as the "moral equivalent of war," a phrase that was widely ridiculed (along with Carter himself, who wore a cardigan while delivering his speech, to underscore the need to turn down the thermostat).

Carter was prescient. We are going to have to reduce our carbon footprint rapidly, and we can do that only by limiting the amount of fossil fuels released into the atmosphere. But what is the most effective — and least painful — way to achieve that goal? Each time we drive a car, use electricity generated by a coal-fired plant, or heat our homes with gas or oil, carbon dioxide and other heat-trapping gases escape into the air. We can use longer-lasting light bulbs, lower the thermostat (and the air conditioning), drive less, and buy more fuel-efficient cars. That will help, and so will switching to cleaner sources of energy. Flying has also emerged as a major carbon don't — with some reason, since airplanes at high altitudes release at least ten times as many greenhouse gases per mile as trains do. Yet neither transportation — which accounts for 15 percent of greenhouse gases — nor industrial activity (another 15 percent) presents the most efficient way to shrink the carbon footprint of the globe.

Just two countries — Indonesia and Brazil — account for about 10 percent of the greenhouse gases released into the atmosphere. Neither possesses the type of heavy industry that can be found in the West or, for that matter, in Russia or India. Still, only the United States and China are responsible for greater levels of emissions. That is because tropical forests in Indonesia and Brazil are disappearing with incredible speed. "It's really very simple," John O. Niles told me. Niles, the chief science and policy officer for the environmental group Carbon Conservation, argues that spending $5 billion a year to prevent deforestation in countries like Indonesia would be one of the best investments the world could ever make. "The value of that land is seen as consisting only of the value of its lumber," he said. "A logging company comes along and offers to strip the forest to make some trivial wooden product or a palm-oil plantation. The governments in these places have no cash. They are sitting on this resource that is doing nothing for their economy. So when a guy says, 'I will give you a few hundred dollars if you let me cut down these trees,' it's not easy to turn your nose up at that. Those are dollars people can spend on schools and hospitals."

The ecological impact of decisions like that are devastating. Decaying trees contribute greatly to increases in the levels of greenhouse gases. Plant life absorbs CO_2. But when forests disappear, the Earth loses one of its two essential carbon sponges (the other is the ocean). The results are visible even from space. Satellite photographs taken over Indonesia and Brazil show thick plumes of smoke rising from the forest. According to the latest figures, deforestation pushes nearly 6 billion tons of CO_2 into the atmosphere every year. That amounts to 30 million acres — an area half the size of the United Kingdom — chopped down each year. Put another way, according to one recent calculation, during the next twenty-four hours the effect of losing forests in Brazil and Indonesia will be the same as if 8 million people boarded airplanes at Heathrow Airport and flew en masse to New York.

"This is the greatest remaining opportunity we have to help address global warming," Niles told me. "It's a no-brainer. People are paying money to go in and destroy those forests. We just have to pay more to prevent that from happening." Niles's group has proposed a trade: "If you save your forest and we can independently audit and verify it, we will calculate the emissions you have saved and pay you for that." The easiest way to finance such a plan, he is convinced, would be to use carbon-trading allowances. Anything that prevents carbon dioxide from entering the atmosphere would have value that could be quantified and traded. Since undisturbed farmland has the same effect as not emitting carbon dioxide at all, people could create allowances by leaving their forests untouched or by planting new trees. (Rainforests are essential to planetary vitality in other ways, too, of course. More than a third of all terrestrial species live in forest canopies. Rising levels of CO_2 there alter the way that forests function, threatening to increase flooding and droughts and epidemics of plant disease. Elevated CO_2 in the forest atmosphere also reduces the quality of the wood in the trees, and that in turn has an impact on the reproduction of flowers, as well as that of birds, bees, and anything else that relies on that ecosystem.)

From both a political and an economic perspective, it would be easier and cheaper to reduce the rate of deforestation than to cut back significantly on air travel. It would also have a far greater impact on climate change and on social welfare in the developing world. Possessing rights to carbon would grant new power to farmers who, for the first time, would be paid to preserve their forests rather than destroy them. Unfortunately, such plans are seen by many people as morally unattractive. "The whole issue is tied up with the misconceived notion of 'carbon colonialism,'" Niles told me. "Some activists do not want the Third World to have to alter their behavior, because the problem was largely caused by us in the West."

Environmental organizations like Carbon Trade Watch say that reducing our carbon footprint will require restructuring our lives, and that before we in the West start urging the developing world to do that, we ought to make some sacrifices; anything else would be the modern equivalent of the medieval practice of buying indulgences as a way of expiating one's sins. "You have to realize that in the end, people are trying to buy their way out of bad behavior," Tony Juniper, the director of Friends of the Earth, told me. "Are we really a society that wants to pay rich people not to fly on private jets or countries not to cut down their trees? Is that what, ultimately, is morally right and equitable?"

Sandor dismisses the question. "Frankly, this debate just makes me want to scream," he told me. "The clock is moving. They are slashing and burning and cutting the forests of the world. It may be a quarter of global warming and we can get the rate to two percent simply by inventing a preservation credit and making that forest have value in other ways. Who loses when we do that?

"People tell me, well, these are bad guys, and corporate guys who just want to buy the right to pollute are bad, too, and we should not be giving them incentives to stop. But we need to address the problems that exist, not drown in fear or lose ourselves in morality. Behavior changes when you offer incentives. If you want to punish people for being bad corporate citizens, you should go to your local church or synagogue and tell God to punish them. Because that is not our problem. Our problem is global warming, and my job is to reduce greenhouse gases at the lowest possible cost. I say solve the problem and deal with the bad guys somewhere else."

The Tesco corporate headquarters are spread across two low-slung, featureless buildings in an unusually dismal part of Hertfordshire, about half an hour north of London. Having inspired many of the discussions about the meaning of our carbon footprint, the company has been criticized by those who question the emphasis on food. As Adrian Williams, the Cranfield agricultural researcher, put it, the company has been "a little bit shocked" by the discovery that its original goal, to label everything, was naive.

The process has indeed been arduous. Tesco has undertaken a vast — and at times lonely — attempt to think about global warming in an entirely new way, and the company shows little sign of pulling back. "We are spending more than a hundred million pounds a year trying to increase our energy efficiency and reduce CO_2 emissions," Katherine Symonds told me. A charismatic woman with an abiding belief that global warming can be addressed rationally, Symonds is the corporation's climate-change manager. "We are trying to find a way to help consumers make choices they really want to make — choices that mean something to them. This is not all about food. We just happen to be in the food business.

"One of our real responsibilities is to say to our customers, 'The most important thing you can do to affect climate change is insulate your house properly,'" she went on. "'Next would be to get double-glazed windows,'" which prevent heat from escaping in the winter. "'Third, everyone should get a new boiler.' We are trying to put this into context, not to say, 'Buy English potatoes.'" Consumers are unlikely to stop shopping. Economies won't stand still, either; those of China and India are expanding so speedily that people often ask whether sacrifices anywhere else can even matter.

"We have to be careful not to rush from denial to despair," John Elkington told me, when I visited him not long ago at his offices at

SustainAbility, the London-based environmental consulting firm he helped found more than two decades ago. He believes there is a danger that people will feel engulfed by the challenge, and ultimately helpless to address it.

"We are in an era of creative destruction," he said. A thin, easygoing man with the look of an Oxford don, Elkington has long been one of the most articulate of those who seek to marry economic prosperity with environmental protection. "What happens when you go into one of these periods is that before you get to the point of reconstruction, things have to fall apart. Detroit will fall apart. I think Ford" — a company that Elkington has advised for years — "will fall apart. They have just made too many bets on the wrong things. A bunch of the institutions that we rely on currently will, to some degree, decompose. I believe that much of what we count as democratic politics today will fall apart, because we are simply not going to be able to deal with the scale of change that we are about to face. It will profoundly disable much of the current political class."

He sat back and smiled softly. He didn't look worried. "I wrote my first report on climate change in 1978, for Herman Kahn, at the Hudson Institute," he explained. "He did not at all like what I was saying, and he told me, 'The trouble with you environmentalists is that you see a problem coming and you slam your foot on the brakes and try and steer away from the chasm. The problem is that it often doesn't work. Maybe the thing to do is jam your foot on the pedal and see if you can just jump across.' At the time, I thought he was crazy, but as I get older I realize what he was talking about. The whole green movement in technology is in that space. It is an attempt to jump across the chasm."

PATRICK SYMMES

Red Is the New Green

FROM *Outside*

THE KEY TO A GOOD TRIP is of course a bad start. So things began looking up for me in Cuba late on my fourth night, when I was ambushed by two young men. They smashed me against the door of my guesthouse, grabbed my wallet, and took off sprinting.

I went after them. Nobody has a gun in Havana, or practically anything sharper than a butter knife, so this isn't quite as stupid as it sounds. I ran hard, chasing them through the quiet city center, screaming for help. After a block I kicked off my flip-flops and started closing the gap, but it was too late. The boys leaped onto their escape vehicle, a Flying Pigeon, one of the 1.2 million bicycles imported from China in the early 1990s, part of Cuba's move away from fossil fuels toward nonpolluting transport. The Flying Pigeon weighs about fifty-five pounds, but we were headed downhill now, past the University of Havana, and with one boy pedaling and the other standing on the pegs, they squeaked into the night.

For the next two weeks, things just kept getting better. I went scuba diving and blew out my eardrum. A guy dropped an air tank on my foot. The bloody four-inch scrape on my left arm, souvenir of the mugging, turned green and filled with pus. I was bitten by a crab, rafts of mosquitoes, and two dozen sand mites. Although I'd come here to measure Cuba's environmental situation in the twilight of Fidel's reign, many of the greens I was looking for were in hiding, or jail, or exile. People kept whispering that everything was a lie. I looked up an old friend; he'd become an alcoholic. It was June, and even the weather was bad: windless, humid, and blistering.

Ah, Cuba, *mi amor.* Dreams are duty-free, imported with our carry-on bags; disillusionment is the national export. We bring the paradise, Cuba supplies the music and mojitos, the good scuba diving, the confounding moral examples, the surprisingly intact ecosystems, and — ouch! — a quick and bloody mugging, all accomplished with amazingly bad equipment.

This was my lucky thirteenth visit. Things may start off badly, but in Cuba the future is always glorious. No matter how much they beat on me, I will say it again: everything is splendid in Cuba. Don't believe otherwise, no matter what I tell you.

There is only one country on Earth that is truly, deeply, accidentally green. In 2006 the World Wildlife Fund (WWF) did the global math, cross-indexing social factors like education and life expectancy with each population's ecological footprint and global biocapacity. The poor countries were huddled on the left, underdeveloped. The rich soared up to the right, overconsuming. Out of the 150 nations studied, only here, in the rigidly ruled kingdom of Dr. Castro, were human beings developing at a statistically sustainable rate.

WWF isn't alone in concluding that Cuba is doing something right. The United Nations lists it as one of the only countries in the Caribbean that have stopped and even reversed deforestation, with 22 percent of the island covered in everything from palms and pines to mangroves and ceiba trees. Fidel Castro has declared himself the island's chief ecologist, intervening in everything from the design of lobster boats (he added live wells, to ensure that juveniles return to the sea still kicking) to the retrofitting of sugar mills (to run on bagasse, the biomass left behind by the sugar harvest). He ordered the planting of 348 million trees in the 1960s, and as recently as 2007 Cuba claimed it would put 135 million more in the ground. At the Rio Earth Summit, in 1992, Fidel received rapturous applause as a prophet of low-consumption ethics, even as George H. W. Bush was criticized for ignoring both him and the resulting treaties.

Since then Cuba's national assembly has enshrined sustainable development in the constitution, designated 20 percent of the country for conservation, and organized an island-wide drive to install fluorescent light bulbs and rip out old electricity-sucking refrigerators. When Castro learned that spearfishing was damaging the Jardines de la Reina national park, an 830-square-mile archipelago in Cuba's south, he simply banned it — even though spearfishing there was his favorite relaxation.

There can be no question that something genuinely important is now at stake in Cuba. If you think you know the Caribbean, think again. Cuba *is* the Caribbean: it contains nearly half the landmass (ten times that of Jamaica) and a third of the population (11 million people) in that round sea. And in a region stressed by development, habitat loss, overfishing, and low environmental standards, Cuba retains the richest biodiversity in the Caribbean, with perhaps half of her 20,808 known terrestrial species found nowhere else. Its seas are like no other: the northern coast stretches in a bow almost a thousand miles long, with crystalline waters, blue holes, and some thousand barrier islands barely fathomed since Hemingway sent his U-boat hunters stalking their channels in *Islands in the Stream*. In the south, nutrients well up from the 25,000-foot-deep Cayman Trench, enriching a seventy-mile-wide continental shelf. As much as 50 percent of Cuba's southern coast is healthy red and black mangrove forest, the largest such forest — and fish nursery — in the Caribbean. Its 2,200-square-mile Zapata Swamp is the Caribbean's largest wetland, home to nine hundred plant species, and its reefs are intact to a standard unknown elsewhere in the region.

Hold on a minute: Cuba? For as long as I've been reporting on the island — fifteen years now — I've listened skeptically to declarations about Cuba's ecological achievements, which seemed to defy the brown reality I saw on the ground. Eurogreen Web sites trumpet wind farms; I found the beaches of the north coast stained with tar from a low-tech, mismanaged oil industry. The head of the Center for Cuban Studies, in New York, told me how organic gardening was greening Havana; in the countryside I found crude state agriculture despoiling the earth. The United Nations gave biosphere status to a wildlife reserve in Pinar del Río province surrounding an "ecovillage" called Las Terrazas; this turned out to be mostly some phony thatched huts and a sandwich shop used by tour buses.

If the present is dubious, the future may be worse. After nearly five decades of Castro and a collective twelve years of Bush, regime change is the new reality. Now eighty-one, Fidel is on life support;

his seventy-six-year-old brother, Raúl, is in the wings; and presiden-
tial candidates from Barack Obama to John McCain have spoken
cautiously of a new beginning with Cuba. Lifting the U.S. embargo
would send millions of American tourists and billions of investment dollars flooding into Cuba, turning pristine coastline into —
well, think Ayn Rand with a cement mixer. Cubans today are hungry and cash-starved but tightly controlled; a sudden collapse of
the country's political system would unleash them — and foreign
investors — on the forests and seas.

In Cuba facile lessons are abundant (Michael Moore, call your
doctor!), but even a jaded visitor like me has to acknowledge that a
particular mix of Cuban factors — utter economic incompetence,
visionary green policies, and a dash of red brutality — has con-
spired to deliver the big island into the twenty-first century with an
almost nineteenth-century set of natural assets. This success comes
with a question: is it possible that Cuba, a deliberate refutation
of the consumer-crazed, gas-guzzling, climate-warming American
way of life, might someday offer us genuine alternatives? If Super-
tanker America strikes a reef, wouldn't we need some quick lessons
from a neighbor about urban gardening, low-impact living, and
fitting two guys onto one bicycle?

Aside from being the worst divemaster I've ever encountered, the
guide who led us into the deep waters off Cuba's western tip was
also a *bobo*. A *bobo* is a cherished Cuban type, someone who is deliberately, defiantly, entertainingly full of it. In this case the divemaster was a laughing complainer who evaded all our questions,
skipped the safety briefing entirely so he could smoke a cigarette,
didn't bother telling us where we were going, and said of the marine life we might encounter under the surface, "Fishes we will discuss later." He also looked ridiculous in his pseudo-Speedo.

But the water . . . ah, the water. River sediments make diving
off the beaches of eastern Cuba unimpressive, but here in the
arid west, facing straight toward Cozumel at a small resort called
María la Gorda, upwellings of clean water and marine life from the
Cayman Trench create some of the best diving in the Caribbean.
Every ripple of sand seemed just beyond reach, even though it was
thirty feet away.

I quickly found myself in a bubble bath. My vest, regulator, and
backup regulator were all streaming air. "All equipment leaks in

Cuba, always," the diver next to me said. He was a blond New Zealander in his twenties, named Ryan. He suggested we team up — something the divemaster hadn't bothered with — and we plunged downward, following the other divers as the shelf fell away into one of Cuba's steep walls, or veriles, an express ride to the bottom of the Caribbean.

The divemaster led us to a narrow crack in the sand, a kind of crevasse in the sea floor, and without looking back he and the others plunged in, with Ryan and me a distant last. The descent was almost vertical, the canyon walls tight and lined with long trees of black coral, one of the slowest-growing, most endangered of all corals. He hadn't mentioned that we were heading right into one of Cuba's most famous black-coral formations, thousands of years of creeping progress now vulnerable to any flailing goofball in a wetsuit.

This old-growth forest seemed like an argument for Cuba's conservation policies — the global trade in black coral is restricted under Schedule II of the CITES treaty, precisely because it's been stripped from places like the Cayman Islands and the Yucatán. But back in Havana, I found pearls of the coral selling in the gift shop of the Havana Libre hotel for a dollar apiece. Cuba's government has allowed itself to do many things - - like netting dolphins and selling them to global water parks, as it did in the 1990s — that would be condemned if done by private interests.

I managed to slide out of the canyon without crushing anything, but just as I came into open water, my left ear blew. I twisted, gasped, and forgot my buoyancy. I drifted down to one hundred feet, alone but for Ryan, watching from afar. The pain was taking me away from things — too deep, too unsafe, breathing too fast (oh, panic!) — but I closed my eyes and managed to correct my depth, following Ryan slowly along the wall, past ten-foot-long pencils of coral, raging purple wrasses, and queen angelfish. An octopus peeked out from below a rock; huge coral jugs littered the wall, a healthy reef buzzing in an emerald-and-blue kingdom.

By the time we made it to the boat, the divemaster was already on the bridge, smoking. We never got a discussion of "the fishes," and back on land, when I asked his name, he stalked off without comment. Accountability is not the Cuban way. As if to emphasize the point, a careless crewman slammed an oxygen tank onto my foot and grinned helplessly under my barrage of curses.

There is a lot of talk about what will happen when American tourists can visit Cuba, but the Europeans and Canadians are already here, 2.3 million of them a year, pouring roughly $2 billion into the island, according to *The Economist*. Over the past ten years, Cuba has doubled the number of hotel rooms, to 50,000, often bypassing environmental concerns in the process. On the hotel-strewn Varadero peninsula, two hours east of Havana, reserves of coconut palms have been bulldozed to accommodate hotels. On the north coast, in 1988, the government rammed a seventeen-mile causeway through shallow bays to the pristine barrier island of Cayo Coco and then built a string of resorts with 3,000 hotel rooms. Dozens of species of birds and fish were damaged severely; across Cuba, biologists, archaeologists, fishermen, and dissidents told me that Cayo Coco was the spark for their green activism, the first time they saw the environment as a political problem. Today the causeway is dotted with police posts that turn away ordinary Cubans, the ultimate example of what Frommer's calls the "apartheid-like tourist sites" where foreigners languish at a remove from the island's reality.

María la Gorda — Fat Maria — was isolated by geography, price, policy. There were no Cuban guests at the little resort, which had a decent beach, nice cabins, and great sunsets. But the food was pathetic, a steam-table array of hot-dog salads, chicken à la defector, and oily fish, much of it recycled at breakfast. This is typical of Cuban resorts, in my experience at least, and explains why Cuba enjoys one of the lowest return rates in international tourism. Everything from aircraft landing fees, among the highest in the world, to a special tourist peso (expensive on the island and worthless off it) makes travel outside the resorts difficult and slow.

Ryan sounded disillusioned. "We spent four days in Havana," he told me as we washed our equipment, "and we reckon that was four days too many." He imitated the cigar vendors and would-be pimps: *"He-joe my fren' my fren' you wanna Habana cigar girlfren' mulatta negra rubia."* He and his friends had liked Viñales, the tranquil tobacco country of the west, but this was the only spot they could love. The water was warmer than in New Zealand, and there were no hustlers, bureaucrats, or even Cubans under the waves.

Tourism in Cuba is often run by the military. María la Gorda is operated by Gaviota S.A., Cuba's largest tourism company, itself a wholly owned subsidiary of the Revolutionary Armed Forces. The pilot of one sightseeing helicopter told me he'd learned his trade in a Soviet gunship in Afghanistan; my domestic flights were all on Gaviota planes, painted in bright yellows and tropical greens but listed on airport monitors as FAR, Fuerzas Armadas Revolucionarias. Through Gaviota the military offers fishing expeditions, diving excursions, and trips to thermal baths, and all over Cuba, TransGaviota taxis can whisk you from Gaviota restaurants to Gaviota hotels stocked with Gaviota Hoteles body lotion.

At the bar of one such military-tourism complex, the Hotel Nacional, in Havana, I ran into Wayne Smith, a bearded ex-diplomat who was Jimmy Carter's man in Havana, heading the U.S. Interests Section, the closest thing our government has to an embassy here. Wearing a guayabera shirt, he was sitting in a wicker armchair on the back terrace of this stone palace. Musicians strolled past, as in the days when Al Capone, Winston Churchill, and a pile of movie stars slept here.

Smith now directs the Cuba program for the Center for International Policy in Washington, D.C. He'd known Castro for decades, so I asked if El Comandante was really responsible for Cuba's green successes. "Who knows?" he said. "Fidel is a diver. But it is a very mixed record. Cuba has very strong policies in favor of the environment, but on an institutional level, when it is a question of development versus ecology, it slips through the cracks. What's it look like to you?"

The usual, I said: visionary declarations and then chaos. He grunted. The Cuban environment was just like this hotel: museum more than refuge, protected and neglected, the state skimming off the top and excluding Cubans at the door.

David Guggenheim, an adviser to the Harte Research Institute for Gulf of Mexico Studies, recently told the *New York Times* that diving in Cuba was "like going back in time fifty years," and I'd assumed as much at María la Gorda. But Smith wasn't having it: in the 1950s, he said, "You never went out without seeing schools of barracuda, sharks, clouds of fish." Now, he said, "most of it's gone. People need the fish, the protein."

That night I tasted sustainable ocean policy for myself at La Divina Pastora, a restaurant in an old Spanish fortress across the harbor. Cuba uses holistic techniques to breed up a healthy lobster population with habitat improvements, Fidel's live wells, and catch limits enforced with a ruthlessness that only a dictator could achieve. (The private sale of a lobster is a crime in Cuba; police hunt down black-market lobstermen as if they were crack dealers.) This allows Cuba to export frozen seafood to Europe and Caribbean resorts but also benefits Americans: since lobster larvae drift up the Gulf Stream, Cuba's management actually helps restore American waters.

The restaurant was run by Gaviota. Dressed in a Hawaiian shirt, the maître d' introduced himself as "Revolutionary Armed Forces Lieutenant Colonel Gerardo Tur." Gaviota, he conceded, had "a close, very direct relationship" with the military. He himself had been seconded to the front lines of the tourism struggle and bragged of having "the best wine list in Cuba." I didn't test him, ordering a daiquiri instead, and then, sitting behind a line of nine iron cannons, ate fresh-picked lobster meat with lime mayonnaise.

Behind me was El Morro, the lighthouse-in-a-castle that is Cuba's most famous landmark. Down the sea wall I could see the U.S. Interests Section, glowing with anti-Castro propaganda and flanked by retaliatory billboards of Uncle Sam having his butt kicked by sexy Cubans.

At 9 P.M. the harbor gun went off, just as it has for centuries. It used to mean the port was closed for the night. These days it is a starter pistol for the debaucheries to come.

A few mornings later, at 4:20 A.M., I boarded an Ilyushin turboprop, the finest 1960s technology, for a smoky, rumbling flight to the eastern province of Holguín. A thundering salsa band greeted our 7 A.M. arrival, and the other passengers — European tourists — boarded buses to local beach resorts, said to be the most luxurious in Cuba. In a dusty parking lot, I negotiated for the front seat in an old Toyota jeep headed to Moa Bay, said to be the most polluted.

It is also one of the most remote. For ten dollars — three times what the five Cubans in back were paying — I got the front seat during a three-hour drive east. We climbed into the Sierra del Cristal, one of the most pristine places in Cuba, land of small streams, hidden groves of palm trees, tight valleys alternating with long vistas. Just four years ago a farmer discovered an "extinct" mammal, the groundhog-size almiqui, rooting in his crops.

The mountains gave way to the farming plains where Fidel himself grew up. Now it was more Marx Brothers than Marx: abandoned farms, bulldozed trees, cattle ranches with more cowboys than cows, and a horse-breeding station that, the driver assured me, had only one horse. Cuba in a nutshell. By the time I made it to Moa at midday, I'd traded the Toyota for a Nash Rambler with a squealing piglet lashed to the bumper. In the back of a bicycle rickshaw, I went looking for the underground environmentalists.

This was slow going at first. The chain on this three-wheeled contraption kept slipping loose, and the driver would pedal forward, then remount the chain by pedaling backward. We lurched down Moa's main street in this overtly symbolic manner ("That's Cuba!" some *bobo* shouted) to a humble wood house tilted over a dirty creek, the home of Silverio Herrera Acosta, an asthmatic fifty-five-year-old photographer. With him was Francisco Hernandez Gomez, a young looking thirty-six-year-old activist. These two were it, their own tiny movement, without even a telephone yet subject to repression and arrest.

For all its green gains, Cuba is still a police state, as the island's few independent environmentalists are constantly reminded. Opposition ecologists and amateur greens who try to organize or protest can face an escalating menu of retaliation: career coldness and lack of promotions, followed by lecturing, then threats, then informal and formal detentions, all the way up to serious jail time. The founders of Naturpaz, a tiny, illegal group in Havana, have been arrested repeatedly for advertising the state of the filthy streams and streets in their slums. Cuba's real environmental policy, Silverio told me, is "Shut your mouth." Fidel's conservation initiatives are like "a woman who puts on makeup but doesn't bathe."

Silverio and Francisco had become activists only because Moa required it: an American company, Freeport Sulphur, built a refinery here in the 1950s, when this area was still wilderness; now Cuba operates three Soviet-built smelters — known collectively as the Che Guevara complex — and leases the American refinery to Sherritt International Corporation of Canada (which says that it does its heavy refining elsewhere). In a fit of central planning, the town of 65,000 workers was built directly downwind, and the result, Silverio said, is an epidemic of asthma and one of Cuba's highest lung-cancer rates. ("Even teenagers get it," he said.) He wiped a finger across his glass coffee table, bringing up a black smear. "I cleaned this yesterday," he said. "It's a residue of hydrochloric acid, caustic soda, ammonia, and others."

Francisco tried going through the proper channels. He wrote a letter; a government team came and looked at the nickel dust; he never heard another word. "You can protest to the government," Francisco said, "but it's just for your own pleasure."

In 2006 he started to organize a survey of asthmatic children, but the police came to his house four times, he says, warning him to stop. That June, he claims, he was attacked and beaten by government sympathizers and, when he tried to start the survey, summoned to police headquarters, where he says he was kicked in the head, back, and kidneys. Almost a year later, he pulled up his Puma T-shirt and showed me a long white scar.

Francisco and I were crossing an intersection when he shoved me into the bushes. A motorcycle cruised past slowly, the rider wearing an orange helmet.

"That's the guy who watches us," Francisco explained. "We aren't afraid of jail," he added, with more resignation than bravado. "If we have to go to jail, we'll go."

We went to the refineries instead. We hired a thundering 1950s wagon and rode through a landscape out of Road Runner v. Coyote, with red earth, yellow sulfur deposits, and clouds of billowing steam. Earthen berms made it difficult to see anything, but later Eudel Cepero, an exiled Cuban environmentalist who was teaching at Florida International University, steered me to the coordinates on Google Earth. From outer space I inspected the pockmarked minescape, its gigantic slurry pools spilling plumes into the sea. This too is Castro's environmental legacy.

I had to get out of Moa. All three hotels rejected me, the daily buses had gone, and there were no cars for hire. At nightfall Francisco walked me to the highway outside town, where I waited until 11 P.M., one of thirty people in the dark hoping to flag down a ride. Finally, the patient Francisco suggested I sleep illegally at his house, even though "the captain will send men to beat you."

We walked home — zero carbon emissions — and before midnight arrived at the three-room house he shared with his wife, brother, niece, and mother. The mother sat in the front yard in a rusty dentist's chair, smoking a cigar. I took a Cuban shower — a bucket of water, a cup, and a rag — and lay down on the family's best bed, a coil of taut ropes punctured by broken springs. Mosquitoes, breezes, and finally rain blew through the boarded walls.

Screw the resorts anyway: this was the Cuba I loved, generous and striving, principled and poisoned, hotheaded, barefoot, and stubborn, stirringly alive. No amount of cynicism (or accuracy) can overwhelm that romantic mix of Cuban genius and chaos, mythology and self-contradiction.

I lay on Francisco's rope mattress for the next hour, too exhausted to sleep, counting glass bottles. Like all Cubans, his family carefully washed and reused them, and there were forty-two stacked in the corner. Cuba is less a sustainable society than a preconsumer one. Everyone walks or rides everywhere. Dumpsters are empty, because virtually everything can be fixed, sold, or traded, or fed to pigs. Paper is rounded up by old men on the state payroll. Everything from kitchen greens and eggs to pork rinds and table fish is gathered so slowly and so seasonally that it would make even a devout locavore swim for Key West in frustration.

Only necessity has made Cuba green, which may be the island's real lesson: No transportation. No shopping. No advertising. No energy. No waste, no fat, and no gristle, no conspicuous consumption, and not much inconspicuous consumption either. Noneconomy is green economy.

Eventually Francisco brought me a nightcap that knocked me out. It was a glass of water, reeking of sulfur.

Fidel Castro may go to hell for the things he has done, but if he gets into heaven it might be for what he hasn't done — what has not been despoiled, destroyed, polluted, or paved. After that earbusting visit to María la Gorda, I had taken my rental car even farther west, out the last, long, potholed road in Cuba, to the westernmost point in the West Indies, Cabo San Antonio. Here was one of the Caribbean's last refuges for a true dinosaur, *Chelonia mydas,* the oceangoing green turtle.

The land was flat, scrubby, and bitter, and only tiny iguanas — "shore puppies" in Cuban parlance — moved through the brush. After hours of searching, I spotted not a turtle but a lean, sunburned Cuban in a tight swimsuit and a cowboy hat. He ambled up from the beach, as surprised to see anyone here as I was.

Rolando Díaz is a forty-two-year-old wildlife technician; the government deploys researchers like him and students from the University of Havana to camp out in fifteen-day stints guarding eight beaches where the green turtles breed. He showed me his camp: a Eureka tent in the sea grapes, an AM radio, and a cistern for water. "Here we are trying to save turtles," Rolando said, "so my grandchildren can see them. We've seen as many as one hundred and thirty turtles on this little beach, but there aren't many this year. Something is wrong. Normally they are here by now."

The biggest threat, he said, was not development or predators but ordinary Cubans. Once, when technicians missed a single fifteen-day stint, hungry locals ate about seventy turtles. "We come here from Havana and tell people, 'Don't eat turtles, they are almost gone,'" Rolando explained. "And they say, 'So are we.' The majority here live on turtles. It is hard to do conservation in a poor country. It's hard to be an ecologist in Cuba." His salary was 350 pesos a month, or $13. His own brother, head of the turtle program, had gone to a conference in Greece and never come home.

I stood the beach vigil all that long night with Rolando. The females come ashore most nights in the summer, dragging their dinosaur carapaces to the tree line and scooping out huge bunkers to bury a hundred or more eggs. One mother had come up last night, and Rolando showed me her four-foot-wide path through the sand and the stick he used to mark the new nest. Aside from humans, dogs, feral pigs, gulls, rats, and even crabs would root up the eggs.

At 1:30 A.M., the night clouded over, hiding the moon. The turtles preferred this total darkness, but it made every rock in the surf look like a carapace, and clouds of mosquitoes emerged to torture me. Before retiring, Rolando warned me not to lie down, but I did, and I discovered why: the sand was infested with *jejere,* mites that bit me until blood ran down my ankles. From 2:30 to 3:30 I tried napping on a smooth driftwood plank, then awoke with a yelp when a red crab bit my toe.

Rolando emerged from his tent again, and we sat listening to the waves break. He had the eloquence of a man who has been waiting for a conversation for fifteen days. The ocean was "my friend the sea. It is beautiful above and below. That's my world, the beautiful sea. If I could be reborn, I'd be a fish, a whale, even a shark. I love them all."

"How much would tourists pay for that?" Rolando asked. "To have a crab touch their toe? Fifty dollars? Everyone wants nature. They could come here and see turtles and live in tents . . ."

After a while he went back into the Eureka, but I stayed, walking the empty beach, hours without light, traffic, people, boats, or anything but the sea and its mysteries, above and below.

I never saw a turtle. At 5:15 A.M., when the first crack of purple appeared in the east, I went to my little rental car, twisted across the front seat, and went to sleep.

Everything will be better tomorrow.

GARY WOLF

Stayin' Alive

FROM *Wired*

WITH HIS PERFECT POSTURE and narrow black glasses, he would look at home in an old documentary about Cape Canaveral, but his mission is bolder than any mere voyage into space. He is attempting to travel across a frontier in time, to pass through the border between our era and a future so different as to be unrecognizable. He calls this border the singularity. Kurzweil is sixty, but he intends to be no more than forty when the singularity arrives.

Kurzweil's notion of a singularity is taken from cosmology, in which it signifies a border in spacetime beyond which normal rules of measurement do not apply (the edge of a black hole, for example). The word was first used to describe a crucial moment in the evolution of humanity by the great mathematician John von Neumann. One day in the 1950s, while talking with his colleague Stanislaw Ulam, von Neumann began discussing the ever-accelerating pace of technological change, which, he said, "gives the appearance of approaching some essential singularity in the history of the race beyond which human affairs as we know them could not continue."

Many years later, this idea was picked up by another mathematician, the professor and science fiction writer Vernor Vinge, who added an additional twist. Vinge linked the singularity directly with improvements in computer hardware. This put the future on a schedule. He could look at how quickly computers were improving and make an educated guess about when the singularity would arrive. "Within 30 years, we will have the technological means to create superhuman intelligence," Vinge wrote at the beginning of his

1993 essay "The Coming Technological Singularity: How to Survive in the Post-Human Era." "Shortly after, the human era will be ended." According to Vinge, superintelligent machines will take charge of their own evolution, creating ever smarter successors. Humans will become bystanders in history, too dull in comparison with their devices to make any decisions that matter.

Kurzweil transformed the singularity from an interesting speculation into a social movement. His best-selling books *The Age of Spiritual Machines* and *The Singularity Is Near* cover everything from unsolved problems in neuroscience to the question of whether intelligent machines should have legal rights. But the crucial thing that Kurzweil did was to make the end of the human era seem actionable: he argues that while artificial intelligence will render *biological* humans obsolete, it will not make human consciousness irrelevant. The first AIs will be created, he says, as add-ons to human intelligence, modeled on our actual brains and used to extend our human reach. AIs will help us see and hear better. They will give us better memories and help us fight disease. Eventually, AIs will allow us to conquer death itself. The singularity won't destroy us, Kurzweil says. Instead, it will immortalize us.

There are singularity conferences now, and singularity journals. There has been a congressional report about confronting the challenges of the singularity, and late last year there was a meeting at the NASA Ames Research Center to explore the establishment of a singularity university. The meeting was called by Peter Diamandis, who established the X Prize. Attendees included senior government researchers from NASA, a noted Silicon Valley venture capitalist, a pioneer of private space exploration, and two computer scientists from Google.

At this meeting, there was some discussion about whether this university should avoid the provocative term *singularity*, with its cosmic connotations, and use a more ordinary phrase, like *accelerating change*. Kurzweil argued strongly against backing off. He is confident that the word will take hold as more and more of his astounding predictions come true.

Kurzweil does not believe in half measures. He takes 180 to 210 vitamin and mineral supplements a day, so many that he doesn't have time to organize them all himself. So he's hired a pill wrangler, who takes them out of their bottles and sorts them into daily doses, which he carries everywhere in plastic bags. Kurzweil also spends one day a week at a medical clinic, receiving intravenous longevity treatments. The reason for his focus on optimal health should be obvious: if the singularity is going to render humans immortal by the middle of this century, it would be a shame to die in the interim. To perish of a heart attack just before the singularity occurred would not only be sad for all the ordinary reasons, it would also be tragically bad luck, like being the last soldier shot down on the Western Front moments before the armistice was proclaimed.

In his childhood, Kurzweil was a technical prodigy. Before he turned thirteen, he'd fashioned telephone relays into a calculating device that could find square roots. At fourteen he wrote software that analyzed statistical deviance; the program was distributed as standard equipment with the new IBM 1620. As a teenager, he cofounded a business that matched high school students with colleges based on computer evaluation of a mail-in questionnaire. He sold the company to Harcourt, Brace & World in 1968 for $100,000 plus royalties and had his first small fortune while still an undergraduate at MIT.

Though Kurzweil was young, it would have been a poor bet to issue him life insurance using standard actuarial tables. He has unlucky genes: his father died of heart disease at fifty-eight, his grandfather in his early forties. He himself was diagnosed with high cholesterol and incipient type 2 diabetes — both considered to be significant risk factors for early death — when only thirty-five. He felt his bad luck as a cloud hanging over his life.

Still, the inventor squeezed a lot of achievement out of these early years. In his twenties, he tackled a science fiction type of problem: teaching computers to decipher words on a page and then read them back aloud. At the time, common wisdom held that computers were too slow and too expensive to master printed text in all its forms, at least in a way that was commercially viable.

But Kurzweil had a special confidence that grew from a habit of mind he'd been cultivating for years: he thought exponentially. To illustrate what this means, consider the following quiz: 2, 4, ?, ?.

What are the missing numbers? Many people will say 6 and 8. This suggests a linear function. But some will say the missing numbers are 8 and 16. This suggests an exponential function. (Of course, both answers are correct. This is a test of thinking style, not math skills.)

Human minds have a lot of practice with linear patterns. If we set out on a walk, the time it takes will vary linearly with the distance we're going. If we bill by the hour, our income increases linearly with the number of hours we work. Exponential change is also common, but it's harder to see. Financial advisers like to tantalize us by explaining how a tiny investment can grow into a startling sum through the exponential magic of compound interest. But it's psychologically difficult to heed their advice. For years an interest-bearing account increases by depressingly tiny amounts. Then, in the last moment, it seems to jump. Exponential growth is unintuitive, because it can be imperceptible for a long time and then move shockingly fast. It takes training and experience, and perhaps a certain analytical coolness, to trust in exponential curves whose effects cannot be easily perceived.

Moore's law — the observation by Intel cofounder Gordon Moore that the number of transistors on an integrated circuit doubles roughly every eighteen months — is another example of exponential change. For people like Kurzweil, it is the key example, because Moore's law and its many derivatives suggest that just about any limit on computing power today will be overcome in short order. While Kurzweil was working on his reading machine, computers were improving, and they were indeed improving exponentially. The payoff came on January 13, 1976, when Walter Cronkite's famous sign-off — "and that's the way it is" — was read not by the anchorman but by the synthetic voice of a Kurzweil Reading Machine. Stevie Wonder was the first customer.

The original reader was the size of a washing machine. It read slowly and cost $50,000. One day late last year, as a winter storm broke across New England, I stood in Kurzweil's small office suite in suburban Boston, playing with the latest version. I hefted it in my hand, stuck it in my pocket, pulled it out again, then raised it above a book flopped open on the table. A bright light flashed, and a voice began reading aloud. The angle of the book, the curve of its pages, the uneven shadows — none of that was a problem. The mechanical voice picked up from the numerals on the upper left corner — . . . *four hundred ten. The singularity is near. The continued oppor-*

tunity to alleviate human distress is one key motivation for continuing technological advancement — and continued down the page in an artificial monotone. Even after three decades of improvement, Kurzweil's reader is a dull companion. It expresses no emotion. However, it is functionally brilliant to the point of magic. It can handle hundreds of fonts and any size book. It doesn't mind being held at an angle by an unsteady hand. Not only that, it also makes calls: computers have become so fast and small they've nearly disappeared, and the Kurzweil reader is now just software running on a Nokia phone.

In the late 1970s, Kurzweil's character-recognition algorithms were used to scan legal documents and articles from newspapers and magazines. The result was the Lexis and Nexis databases. And a few years later, Kurzweil released speech-recognition software that is the direct ancestor of today's robot customer service agents. Their irritating mistakes taking orders and answering questions would seem to offer convincing evidence that real AI is still many years away. But Kurzweil draws the opposite conclusion. He admits that not everything he has invented works exactly as we might wish. But if you will grant him exponential progress, the fact that we already have virtual robots standing in for retail clerks, and cell phones that read books out loud, is evidence that the world is about to change in even more fantastical ways.

Look at it this way: if the series of numbers in the quiz mentioned earlier is linear and progresses for 100 steps, the final entry is 200. But if progress is exponential, then the final entry is 1,267,650,600,228,229,400,000,000,000,000. Computers will soon be smarter than humans. Nobody has to die.

In a small medical office on the outskirts of Denver, with windows overlooking the dirty snow and the golden arches of a fastfood mini-mall, one of the world's leading longevity physicians, Terry Grossman, works on keeping Ray Kurzweil alive. Kurzweil is not Grossman's only client. The doctor charges $6,000 per appointment, and wealthy singularitarians from all over the world visit him to plan their leap into the future.

Grossman's patient today is Matt Philips, thirty-two, who became independently wealthy when Yahoo bought the Internet advertising company where he worked for four years. A young medical technician is snipping locks of his hair, and another is extracting small vials of blood. Philips is in good shape at the moment, but he is aware that time marches on. "I'm dying slowly. I can't feel it, but I know its happening, little by little, cell by cell," he wrote on his intake questionnaire. Philips has read Kurzweil's books. He is a smart, skeptical person and accepts that the future is not entirely predictable, but he also knows the meaning of upside. At worst, his money buys him new information about his health. At best, it makes him immortal.

"The normal human lifespan is about one hundred and twenty-five years," Grossman tells him. But Philips wasn't born until 1975, so he starts with an advantage. "I think somebody your age, and in your condition, has a reasonable chance of making it across the first bridge," Grossman says.

According to Grossman and other singularitarians, immortality will arrive in stages. First, lifestyle and aggressive anti-aging therapies will allow more people to approach the 125-year limit of the natural human lifespan. This is bridge one. Meanwhile, advanced medical technology will begin to fix some of the underlying biological causes of aging, allowing this natural limit to be surpassed. This is bridge two. Finally, computers will become so powerful that they can model human consciousness. This will permit us to download our personalities into nonbiological substrates. When we cross this third bridge, we become information. And then, as long as we maintain multiple copies of ourselves to protect against a system crash, we won't die.

Kurzweil himself started across the first bridge in 1988. That year he confronted the risk that had been haunting him and began to treat his body like a machine. He read up on the latest nutritional research, adopted the Pritikin diet, cut his fat intake to 10 percent of his calories, lost forty pounds, and cured both his high cholesterol and his incipient diabetes. Kurzweil wrote a book about his experience, *The 10% Solution for a Healthy Life.* But this was only the beginning.

Kurzweil met Grossman at a Foresight Nanotech Institute meeting in 1999, and they became research partners. Their object of investigation was Kurzweil's body. Having cured himself of his most pressing health problems, Kurzweil was interested in adopting the most advanced medical and nutritional technologies, but it wasn't easy to find a doctor willing to tolerate his persistent questions. Grossman was building a new type of practice, focused not on illness but on the pursuit of optimal health and extreme longevity. The two men exchanged thousands of e-mails, sharing speculations about which cutting-edge discoveries could be safely tried.

Though both Grossman and Kurzweil respect science, their approach is necessarily improvisational. If a therapy has some scientific promise and little risk, they'll try it. Kurzweil gets phosphatidylcholine intravenously, on the theory that this will rejuvenate all his body's tissues. He takes DHEA (dehydroepiandrosterone) and testosterone. Both men use special filters to produce alkaline water, which they drink between meals in the hope that negatively charged ions in the water will scavenge free radicals and produce a variety of health benefits. This kind of thing may seem like quackery, especially when promoted by various New Age outfits touting the "pH miracle of living." Kurzweil and Grossman justify it not so much with scientific citations — though they have a few — but with a tinkerer's shrug. "Life is not a randomized, double-blind, placebo-controlled study," Grossman explains. "We don't have that luxury. We are operating with incomplete information. The best we can do is experiment with ourselves."

Obviously, Kurzweil has no plan for retirement. He intends to sustain himself indefinitely through his intelligence, which he hopes will only grow. A few years ago he deployed an automated system for making money on the stock market, called FatKat, which he uses to direct his own hedge fund. He also earns about $1 million a year in speaking fees.

Meanwhile, he tries to safeguard his well-being. As a driver he is cautious. He frequently bicycles through the Boston suburbs, which is good for physical conditioning but also puts his immortality on the line. For most people, such risks blend into the background of life, concealed by a cheerful fatalism that under ordinary conditions we take as a sign of mental health. But of course Kurzweil objects to this fatalism. He wants us to try harder to survive.

His plea is often ignored. Kurzweil has written about the loneliness of being a singularitarian. This may seem an odd complaint, given his large following, but there is something to it. A dozen of his fans may show up in Denver every month to initiate longevity treatments, but many of them, like Matt Philips, are simply hedging their bets. Most health fanatics remain agnostic, at best, on the question of immortality.

Kurzweil predicts that by the early 2030s, most of our fallible internal organs will have been replaced by tiny robots. We'll have "eliminated the heart, lungs, red and white blood cells, platelets, pancreas, thyroid and all the hormone-producing organs, kidneys, bladder, liver, lower esophagus, stomach, small intestines, large intestines, and bowel. What we have left at this point is the skeleton, skin, sex organs, sensory organs, mouth and upper esophagus, and brain."

In outlining these developments, Kurzweil's tone is so calm and confident that he seems to be describing the world as it is today, rather than some distant, barely imaginable future. This is because his prediction falls out cleanly from the equations he's proposed. Knowledge doubles every year, Kurzweil says. He has estimated the number of computations necessary to simulate a human brain. The rest is simple math.

But wait. There may be something wrong. Kurzweil's theory of accelerating change is meant to be a universal law, applicable wherever intelligence is found. It's fine to say that knowledge doubles every year. But then again, what is a year? A year is an astronomical artifact. It is the length of time required by Earth to make one orbit around our unexceptional star. A year is important to our nature, to our biology, to our fantasies and dreams. But it is a strange unit to discover in a general law.

"Doubling every year," I say to Kurzweil, "makes your theory sound like a wish."

He's not thrown off. A year, he replies, is just shorthand. The real equation for accelerating world knowledge is much more complicated than that. In his book, he gives it as

$$W = W_o \exp\left(\frac{c_1 c_2 c_3}{c_4} \, e^{c_4^t} \right).$$

He has examined the evidence and welcomes debate on the minor details. If you accept his basic premise of accelerating growth, he'll yield a little on the date he predicts the singularity will occur. After all, concede accelerating growth, and the exponential fuse is

lit. At the end you get that big bang: an explosion in intelligence
that yields immortal life.

Despite all this, people continue to disbelieve. There is a lively
discussion among experts about the validity of Moore's law. Kurzweil pushes Moore's law back to the dawn of time and forward to
the end of the universe. But many computer scientists and historians of technology wonder if it will last another decade. Some suspect that the acceleration of computing power has already slowed.

There are also philosophical objections. Kurzweil's theory is that
superintelligent computers will necessarily be human, because
they will be modeled on the human brain. But there are other
types of intelligence in the world — for instance, the intelligence
of ant colonies — that are alien to humanity. Grant that a computer, or a network of computers, might awaken. The consciousness of this fabulous AI might remain as incomprehensible to us as
we are to the protozoa.

Other pessimists point out that the brain is more than raw processing power. It also has a certain architecture, a certain design. It
is attached to a specific type of nervous system, it accepts only particular kinds of inputs. Even with better computational speed driving our thoughts, we might still be stuck in a kind of evolutionary
dead end, incapable of radical self-improvement.

And these are the merely intellectual protests Kurzweil receives.
The fundamental cause for loneliness, if you are a prophet of the
singularity, is probably more profound. It stems from the simple
fact that the idea is so strange. "Death has been a ubiquitous, ever-
present facet of human society," says Kurzweil's friend Martine
Rothblatt, the founder of Sirius radio and chair of United Thera-
peutics, a biotech firm on whose board Kurzweil sits. "To tell people you are going to defeat death is like telling people you are going to travel back in time. It has never been done. I would be
surprised if people had a positive reaction."

To press his case, Kurzweil is writing and producing an autobiographical movie with walk-ons by Alan Dershowitz and Tony Rob-
bins. Kurzweil appears in two guises, as himself and as an intelligent computer named Ramona, played by an actress. Ramona has
long been the inventor's virtual alter ego and the expression of his
most personal goals. "Women are more interesting than men," he
says, "and if it's more interesting to be with a woman, it is proba-

bly more interesting to be a woman." He hopes one day to bring Ramona to life and to have genuine human experiences, both with her and as her. Kurzweil has been married for thirty-two years to his wife, Sonya Kurzweil. They have two children — one at Stanford University, one at Harvard Business School. "I don't necessarily only want to be Ramona," he says. "It's not necessarily about gender confusion, it's just about freedom to express yourself."

Kurzweil's movie offers a taste of the drama such a future will bring. Ramona is on a quest to attain full legal rights as a person. She agrees to take a Turing test, the classic proof of artificial intelligence, but although Ramona does her best to masquerade as human, she falls victim to one of the test's subtle flaws: humans have limited intelligence. A computer that appears too smart will fail just as definitively as one that seems too dumb. "She loses because she is too clever!" Kurzweil says.

The inventor's sympathy with his robot heroine is heartfelt. "If you're just very good at doing mathematical theorems and making stock market investments, you're not going to pass the Turing test," Kurzweil acknowledged in 2006 during a public debate with the noted computer scientist David Gelernter. Kurzweil himself is brilliant at math and pretty good at stock market investments. The great benefits of the singularity, for him, do not lie here. "Human emotion is really the cutting edge of human intelligence," he says. "Being funny, expressing a loving sentiment — these are very complex behaviors."

One day, sitting in his office overlooking the suburban parking lot, I ask Kurzweil if being a singularitarian makes him happy. "If you took a poll of primitive man, happiness would be getting a fire to light more easily," he says. "But we've expanded our horizon, and that kind of happiness is now the wrong thing to focus on. Extending our knowledge and casting a wider net of consciousness is the purpose of life." Kurzweil expects that the world will soon be entirely saturated by thought. Even the stones may compute, he says, within two hundred years.

Every day he stays alive brings him closer to this climax in intelligence and to the time when Ramona will be real. Kurzweil is a technical person, but his goal is not technical in this respect. Yes, he wants to become a robot. But the robots of his dreams are complex, funny, loving machines. They are as human as he hopes to be.

Contributors' Notes

Other Notable Science and
Nature Writing of 2008

Contributors' Notes

Wendell Berry lives and farms with his family in Henry County, Kentucky. His most recent book of essays is *The Way of Ignorance.*

John Broome is the White's Professor of Moral Philosophy at the University of Oxford and a Fellow of Corpus Christi College. He was previously a professor of philosophy at the University of St. Andrews and before that a professor of economics at the University of Bristol. His books are *The Microeconomics of Capitalism, Weighing Goods, Counting the Cost of Global Warming, Ethics Out of Economics, Weighing Lives,* and *Rationality Through Reasoning* (forthcoming). This article was written while he held a Major Research Fellowship from the Leverhulme Trust.

Nicholas Carr is the author of *The Big Switch: Rewiring the World, from Edison to Google* and *Does IT Matter?* He has written for *The Atlantic Monthly,* the *New York Times,* the *Guardian, Wired,* and the *Financial Times,* among other periodicals.

Chris Carroll has written about science, the environment, and other topics for *National Geographic* since 2003. He previously worked as a reporter at his hometown paper, the *St. Louis Post-Dispatch,* and at several other newspapers in the United States and Europe. He lives near Baltimore, Maryland, with his wife and two sons.

Andrew Curry is a freelance journalist covering science, history, politics, and culture for a wide range of magazines. He lives in Berlin.

Keay Davidson's books include *Carl Sagan: A Life.* A science journalist and essayist since the late 1970s, he has contributed to numerous magazines, newspapers, encyclopedias, and anthologies, has appeared on many radio and TV broadcasts, and has been a frequent guest lecturer on university campuses. He lives in San Francisco, where he is completing a biography of Thomas S. Kuhn.

Douglas Fox is a freelance science and environmental writer based in northern California. He has reported on stories around the world, from Australia to Antarctica to Papua New Guinea and Mauritius. His work has appeared in *Discover, New Scientist, Popular Mechanics, Science, Conservation, Science News for Kids, Natural History,* and the *Christian Science Monitor.*

Adam Frank received his Ph.D. in physics from the University of Washington. He held postdoctoral and visiting scientist positions at Leiden University and the University of Minnesota. In 1995 he was awarded a Hubble Fellowship. He is now a professor of astrophysics at the University of Rochester, where he studies the birth and death of stars using supercomputer simulations. His first book, *The Constant Fire: Beyond the Science vs. Religion Debate,* was published in January.

Atul Gawande is a general surgeon at Brigham and Women's Hospital in Boston and a staff writer for *The New Yorker.* He is an associate professor at Harvard Medical School and the Harvard School of Public Health. His best-selling books *Complications* and *Better* have been published in more than twenty languages.

David Grimm is a writer at *Science* and the editor of ScienceNOW, *Science*'s daily online news site. He received a Ph.D. in genetics from Yale University before embarking on his science writing career. In addition to *Science,* his work has appeared in *U.S. News and World Report* and *Financial Times.*

Stephen S. Hall writes about molecular biology, genetics, and the intersection of business and biomedicine. In addition to five books, including *Invisible Frontiers* (on the birth of biotechnology) and *Merchants of Immortality* (on the science and politics of stem-cell research), he has written for many magazines, including *The New York Times Magazine* (where he has worked as an editor), *National Geographic, The Atlantic Monthly, The New Yorker, Technology Review,* and *Science.* His next book will be a natural history of human wisdom. He teaches science writing at the Graduate School of Journalism at Columbia University. He lives in Brooklyn, New York, with his wife and two children.

Sue Halpern is the author of five books, including *Can't Remember What I Forgot: The Good News from the Front Lines of Memory Research* and *Four Wings* *and a Prayer: Caught in the Mystery of the Monarch Butterfly,* which is also an award-winning film. She holds a doctorate from Oxford University and is a scholar in residence at Middlebury College.

Walter Isaacson is the president and CEO of the Aspen Institute, a nonpartisan educational and policy studies institute based in Washington, D.C. He has been the chairman and CEO of CNN and the editor of *Time.* He is the author of *Einstein: His Life and Universe, Benjamin Franklin: An American Life,* and *Kissinger: A Biography,* and coauthor of *The Wise Men: Six Friends and the World They Made.* He lives with his wife and daughter in Washington, D.C.

Frederick Kaufman, a contributing editor at *Harper's Magazine,* teaches at the City University of New York's Graduate School of Journalism. His most recent book is *A Short History of the American Stomach.*

Virginia Morell is the author of *Ancestral Passions: The Leakey Family and the Quest for Humankind's Beginnings,* a New York Times Notable Book of the Year (1995); *Blue Nile;* and *Wildlife Wars* (with Richard Leakey). She is a contributing correspondent for *Science* and a regular contributor to *National Geographic.* She is currently working on a book about smart animals and the scientists who study them, to be published by Crown. She lives in Oregon with her husband, two clever cats, and a bossy American Working Farm Collie, Buckaroo.

J. Madeleine Nash is currently a freelance journalist and science writer based in San Francisco. She is a former senior correspondent for *Time* and the author of *El Niño: Unlocking the Secrets of the Master Weather-Maker.* She has won the American Association for the Advancement of Science magazine writing award three times.

Michelle Nijhuis is a contributing editor of *High Country News* and a correspondent for *Orion.* Her reporting on science and the environment has also appeared in *Smithsonian, National Geographic, Audubon,* and the *New York Times,* and she has won several national honors, including an AAAS Science Journalism Award. A lapsed biologist, she was once paid to chase tortoises through the Sonoran Desert (the tortoises usually won). She now lives with her family in rural western Colorado.

Benjamin Phelan's first paying magazine job was as a fact checker at the celebrity gossip magazine *Star.* He has since written for a number of publications, including *Seed, GQ, Harper's Magazine,* the *Oxford American,* and the *Village Voice,* reporting on subjects such as acid techno, antiques, wine, evolutionary theory, nuclear proliferation, books, partying, cryptozoology, politics, creationist paleontology, exotic domestic fruit, and systemic risk in finance. He lives in Louisville, Kentucky.

Virginia Postrel is a contributing editor and columnist for *The Atlantic Monthly* and the author of *The Substance of Style* and *The Future and Its Enemies*. She has been an economics columnist for the *New York Times* and a columnist for *Forbes* and *Forbes ASAP.* From 1989 to 2000 she was the editor of *Reason* magazine. Her personal Web site is at www.dynamist.com, and she is editor in chief of DeepGlamour.net. She is writing a book on glamour for the Free Press.

David Quammen is the author of eleven books, including *The Song of the Dodo* and *The Reluctant Mr. Darwin,* and the editor of a new illustrated edition of Darwin's *Origin of Species.* Quammen is a contributing writer for *National Geographic,* and he writes for *Harper's Magazine* and other publications. He has three times received the National Magazine Award, most recently for "Was Darwin Wrong?" (*National Geographic,* November 2004). He lives in Montana with his wife, Betsy Gaines, a conservationist.

Joshua Roebke was born in a very small town in Ohio and raised in a slightly larger town in Michigan. He studied nuclear physics and Spanish literature at Michigan State University and received a master's degree in theoretical high-energy physics from McGill University in Montreal. In 2005 he dropped out of his Ph.D. program and moved to Brooklyn to become a writer. Now a visiting scholar at the Office for History of Science and Technology at the University of California, Berkeley, he lives and writes in San Francisco. He is at work on his first book, about physics in the twentieth century, titled *The Invisible World.*

Oliver Sacks is a professor of neurology and psychiatry at Columbia University and the university's first Columbia Artist. He is best known for his books of neurological case studies, including *The Man Who Mistook His Wife for a Hat, An Anthropologist on Mars,* and *Musicophilia: Tales of Music and the Brain.* His book *Awakenings* inspired the Oscar-nominated film of the same name as well as the play *A Kind of Alaska,* by Harold Pinter. Sacks is a frequent contributor to *The New Yorker* and *The New York Review of Books.*

Mark A. Smith teaches English at Lock Haven University in central Pennsylvania. He has published short stories, critical essays, and creative nonfiction in *ISLE, Isotope, Arizona Quarterly,* and elsewhere. The essay reprinted here was chosen as the best nature essay of 2008 by the John Burroughs Association.

Michael Specter, who has been on the staff of the *New Yorker* since 1998, writes frequently about science, public health, and the impact of new technologies on society.

Patrick Symmes is the author of two books on Cuba, *Chasing Che* and *The Boys from Dolores,* and has reported from there extensively for *Harper's Magazine,* as well as covering environmental conflicts and sustainable tourism in Latin America and Asia for *GQ, Condé Nast Traveler,* and *Outside,* where he is a contributing editor.

Gary Wolf is a contributing editor at *Wired.*

Other Notable Science and Nature Writing of 2008

SELECTED BY TIM FOLGER

DAVID GESSNER
Loving the West to Death. *OnEarth,* Winter.
DAVID GREISING
The Carbon Frontier. *Bulletin of the Atomic Scientists,* July/August.
GORDON GRICE
Pondering a Parasite. *Discover,* July.
JEROME GROOPMAN
Superbugs. *The New Yorker,* August 11 and 18.
GUY GUGLIOTTA
The Great Human Migration. *Smithsonian,* July.
PETER GWIN
Lost Tribes of the Green Sahara. *National Geographic,* September.

ERIC HAGERMAN
Shock to the System. *Popular Science,* September.
ALEX HALPERIN
Gorillas in Their Midst. *American.com,* December 31.
JOSHUA HAMMER
Trials of a Primatologist. *Smithsonian,* February.
PETER HESSLER
The Road Ahead. *National Geographic,* May.
JIM HOLT
Numbers Guy. *The New Yorker,* March 3.
JOHN HORGAN
War: What Is It Good For? *Discover,* April.

ROBERT D. KAPLAN
Waterworld. *The Atlantic Monthly,* January/February.
MARC KAUFMAN
Is There Alien Life? *Washington Post National Weekly Edition,* July 28–August 3.
VERLYN KLINKENBORG
Our Vanishing Night. *National Geographic,* November.
KEVIN KRAJICK
Joe Blow. *Outside,* January.
LAWRENCE M. KRAUSS AND ROBERT J. SCHERRER
The End of Cosmology? *Scientific American,* March.
ROBERT KUNZIG
Finding the Switch. *Psychology Today,* May/June.

KIM LARSEN
The Killing Machine. *OnEarth,* Winter.
ANDREW LAWLER
Preserving Iraq's Battered Heritage. *Science,* July 4.
DEBORAH LINCE
Human Exposure. *Isotope,* Spring/Summer.
JEFFREY A. LOCKWOOD
(Un)Natural Selection. *Ecotone,* Winter.

LISA MARGONELLI
  Gut Reactions. *The Atlantic Monthly,* September.
JOSEPH MASCO
  Target Audience. *Bulletin of the Atomic Scientists,* July/August.
MICHAEL MASON
  The Vatican's Holy Alliance. *Discover,* September.
STEVE MIRSKY
  Call of the Reviled. *Scientific American,* June.
MICHAEL MOYER
  Breaking Open the Unknown Universe. *Popular Science,* October.

JILL NEIMARK
  Plastic People of the Universe. *Discover,* May.
PETER NICHOLS
  Captain FitzRoy's Protégé. *Ecotone,* Winter.

JEFF O'CONNELL
  Into Dark Waters. *Men's Health,* December.

ERIC A. POWELL
  Do Civilizations Really Collapse? *Archaeology,* March/April.
MATTHEW POWER
  Peak Water. *Wired,* May.
CATHERINE PRICE
  The Anonymity Experiment. *Popular Science,* March.
HEATHER PRINGLE
  Kelp Highways. *Discover,* June.

MARY ROACH
  Almost Human. *National Geographic,* April.

JOHN SEABROOK
  Suffering Souls. *The New Yorker,* November 10.
BILL SHERWONIT
  The Songbird Mystery. *Wildlife Conservation,* November.
ADAM SKOLNICK
  The Doctor, the Dictator, and the Deadly Mosquito. *Men's Health,* March.
BRYAN SMITH
  What If You Woke Up Every Morning to Find Your Memory Erased? *Men's Health,* September.
ERIK SOFGE
  Standing Guard. *Popular Mechanics,* December.
GINGER STRAND
  The Crying Indian. *Orion,* November/December.

KALEE THOMPSON
  Carbon Discredit. *Popular Science,* July.

JIM THORNTON
Your Privates. *Men's Health,* October.
ABIGAIL TUCKER
Farewell to the King? *Smithsonian,* October.

JOHN UPDIKE
Visions of Mars. *National Geographic,* December.

JAMES VLAHOS
Panopticon. *Popular Mechanics,* January.

LOGAN WARD
Fixing the World on $2 a Day. *Popular Mechanics,* August.
PAMELA WEINTRAUB
The Great Imitator. *Psychology Today,* May/June.
KAREN WRIGHT
Consuming Passions. *Psychology Today,* April.

EMILY YOFFE
Well, Excuuuuuse Meee! *Slate,* October 17.

MICHELE ZACKHEIM
Children of a Lesser God. *Discover,* March.